PRAISE FOR

A Tale of Two Climates

"Finally, atmospheric physicist Bill Pekny honestly addresses global warming and climate change in his book *A Tale of Two Climates—One Real, One Imaginary*. The imaginary tale is that the world is coming to an inescapable, flaming end because of human-caused 'climate change.' This is akin to Chicken Little's hysteria of being hit on the head by an acorn and shrieking, 'The sky is falling.'

The real account in his book is the pure, historical, evidence-based science that demonstrates that the sun, water, gravity, and inertia—not humans, chickens, or acorns—are responsible for our mild, controlled, cyclic, and life-sustaining climate.

Bill and Nancy's book is a must-read for those needing to understand the difference between true or false climate science—that would be all of us."

— **DR. JERRY SIMONS,** Doctorate *in* Meteorology and Industrial Electronics *from* Texas A&M University, Semiconductor business owner

"I have known Bill Pekny for a number of years in both a professional and personal capacity. I have always found him to be a careful, dedicated scholar with the highest integrity. In his book, *A Tale of Two Climates*, Bill gets to the heart of the matter by actually looking at what the raw data, both past and present, is telling us about the climate rather than relying on complex, un-validated climate models that have consistently been shown to be wrong. Bill's approach is exactly one I have advocated and that is to look at what the data is telling us. I applaud his work and recommend it to those that really want to begin to understand climate change."

— **DR. STEVEN SATERLIE**, PhD *in* Experimental Physics *from the* University of Wyoming, American Institute of Aeronautics and Astronautics (AIAA) Associate Fellow

"The cover of the book caught my attention immediately. The colors are so vivid and beautiful. The blue marble and flaming Earth has 'movement.' Your eyes can feel the heat of the fire. It made me want to read the book to find out if our Earth will go up in flames and when.

You personalize your book with life lessons—yours. I learned a lot about my cousin and since we grew up in the same area of Chicago, it made me feel as though I was conversing with you. That conversation makes the science more understandable.

The illustrated figures that complement the scientific graphs and charts make each concept more understandable. The chapter takeaways are very concise and easy to read. The book is very interesting and it reminds the reader to use science and not just listen to rhetoric."

— **MARGARET M. WOLOSZYK**, Educator

"A clear and straightforward denunciation of the popular mythology of global warming alarmists. Bill simply presents the truth based on actual scientific evidence."

— **CAPTAIN BRENT HATHAWAY**, Airline Pilot

"In his book, *A Tale of Two Climates,* Bill has taken on the controversial, and often political, subject of climate change. With scientific facts and data, he has unraveled and laid out the truth with straightforward talk, making it understandable to all."

— **ROGER AND JULIE PYLE,** Barber and Beautician extraordinaire

"For so long, I have felt bombarded with dire predictions about the future of our world because of damage to the atmosphere called 'global warming.' Most, if not all, of that information has seemed to me to be either politically or financially motivated. It was very refreshing for me to have the opportunity to read Bill Pekny's excellent book on this subject. He presents very clearly what has happened with our climate in the past and present, and questions the ability of anyone to predict future climate. Bill has no hidden agenda. He just presents the facts in a way that all can understand, and we can draw our own conclusions. Well done!"

— **DR. RICHARD CURTIS**

"This book takes scientific theories and principles and reduces them to concepts readily understood by the layperson. An excellent read."

— **DR. KAYE P. NICHOLS**

"Weather, climate and climate change are 'hot' topics just now, making Bill's book timely and relevant for the current discussion on the world stage. Never before in recorded history has there been so much attention given to, and misinformation about, this topic.

In the face of emotional and heated opinions that create confusing crosscurrents in our society, this studied analysis of our atmosphere and its behavior stands out as unique. It is a cool-headed, fact-based and logical presentation of what is really happening and the basis for it.

Bill has a way of synthesizing and simplifying complex math, physics and science so you and I can grasp the concepts that form the basis of his analysis. We can understand it! And Bill has the experience, training and background to provide credibility to what he says.

I believe this book will become a classic in the field of climate study and will attract a wide audience among those who, without preformed agenda, want to know what's happening out there in the sky and why."

— **HON. STUART T. WALDRIP**, JD *from the* University of Utah, California Superior Court Judge (ret.)

A TALE OF TWO CLIMATES

A TALE OF TWO CLIMATES
ONE REAL ONE IMAGINARY

Bill Pekny

Two Climates, LLC
Midway, Utah

Ordering Information: Quantity orders by U.S. bookstores, wholesalers, educators, corporations, associations, and others. Please contact Two Climates LLC, or visit www.twoclimates.com.

ISBN: 978-1-7349396-0-6 (hardcover)
978-1-7349396-1-3 (Kindle)

Edited by Susan Suffes,
Your Book Is My Business, Inc., New York, New York

Post-edit by Nancy Pekny

Original cover design by Bill Pekny,
Two Climates LLC, Midway, Utah

Cover illustration by Grayraven Industries, Midway, Utah

Cover layout and interior design and typesetting
by Desktop Miracles Inc., Stowe, Vermont

Publisher's Cataloging-In-Publication Data
(Prepared by The Donohue Group, Inc.)

Names: Pekny, Bill, author.
Title: A tale of two climates : one real, one imaginary / Bill Pekny.
Other Titles: Two climates : one real, one imaginary
Description: Midway, Utah : Two Climates, LLC, [2020] | Includes bibliographical references and index.
Identifiers: ISBN 9781734939606 | ISBN 9781734939613 (ebook)
Subjects: LCSH: Climatic changes. | Atmospheric science.
Classification: LCC QC903 .P45 2020 (print) | LCC QC903 (ebook) | DDC 551.6—dc23

First Edition

Printed in the United States of America

To Nancy, my most beautiful and extraordinary wife and eternal partner. You are the wind in my sail and the best thing that has ever happened to me. You have co-authored my life as well as this book with your gentle, down-to-Earth, and common sense influence throughout.

To our sons—Brian, Matt, and Dan—and their families who are among the heirs of this miraculous blue marble we call Planet Earth. Enjoy its magnificence. And my heartfelt thanks for your noteworthy perspectives and inputs to the writing of this book.

Ain't it true, man's a fool?
When it's hot, he wants it cool.
When it's cool, he wants it hot.
Always wanting what it's not.

—Author Unknown

TABLE OF

Contents

LIST OF
Figures

Foreword

BY STEPHEN M. STUDDERT

Stephen Mark Studdert served on the White House staff as Advisor to United States Presidents George Bush, Ronald Reagan, and Gerald Ford.

As part of the so-called "climate justice movement", in recent days millions around the world, mostly emotionally impassioned youth, have marched in a coordinated "global climate strike" where theater frequently upstaged truth, to address "global warming" and dire environmental predictions of the impending catastrophic end of the world.

Yet on the eve of the international event, a newly released scientific retrospective revealed none of the apocalyptic global warming predictions of the past four decades have come true.

So where is the truth? Most of this wild-eyed daily hysteria has been provoked by media noise, anonymous bloggers, scare tactics, supposition, speculation, and even purely selfish economic reasons. All this seems to ignore that our planet historically goes through weather cycles more powerful than human impact.

For most of us who are not trained in climate science, the frightening worldwide clamor about climate change is at best confusing and controversial and at worst downright mystifying.

Finally, a skilled and qualified scientist has given us a clear presentation of the factual issues without an overlay of unreliable political bias or cleverly manipulated poppycock. A welcome and refreshing treatise, *A Tale of Two Climates* is easily readable and readily understandable. As climate activists organize not to inform but to protest, this exceptionally timely and current read is refreshingly unbiased and much needed.

I am neither a meteorologist nor a climate scientist; I am a practitioner of public policy. Having served as a White House advisor to three U.S. Presidents, I've been most often dependent on cumbersome and overly technical scientific literature to inform policy understandings, while at the same time bombarded by predisposed advocates and prejudiced media.

Now, to the "climate issue" there is a lucid, logical, and common-sense source of answers. Without prejudiced sensation, Bill Pekny thoughtfully and rationally supports every statement with documented historical and technical data. What some noisemakers characterize as "settled science", he differentiates between weather versus climate, and with extensive reliable climate-related statistics, figures, and facts makes clear that the conclusions of "global warming science" are indeed not settled. As he empirically documents, the prevalent doomsday rhetoric is more often imaginative guesswork, assumed extrapolation, and biased conjecture.

With an easy to read format sprinkled with inspirational fun, this scientist-author begins with invaluable definitions, most of which are new to the non-scientist reader—which is most of us. Instructive chapter takeaways are helpful, as are the excellent and unambiguous graphics throughout the book. Interlaced between explaining the differences between weather and climate, Bill Pekny shares many warm and often witty personal anecdotes and experiences which add to both our grasp and the author's sincerity.

For anyone wanting to understand what some politicians are calling "the crisis of our time", this book is a must read. *A Tale of Two Climates* raises serious thought-provoking questions and offers rational explanations about this entire issue—usefully accompanied by coherent fact-based solutions.

On a personal note, I've known Bill Pekny for many years. Deeply respected by his scientific peers, he is a man of great personal and intellectual integrity. In a distinguished career marked by impressive scientific accomplishments, *A Tale of Two Climates* may be his most important contribution yet.

AUTHOR'S NOTE

The Most Important Things to Know

Why am I doing this? For some time, I've observed what others are saying about climate. Some insist on leading us down a path of controversy about our atmosphere, based on incomplete information and supposition and myth. So, here I will tell what is true, and what is not, based on researched and authenticated evidence rather than on hearsay, money-motivated maneuvers, unsubstantiated climate theories, and misleading computer models.

Climate science is complicated and it is not thoroughly understood, yet. But, quite a bit of it is understood well, and you can get a look at that here. The purpose of this book is to share what I've observed and examined over my lifetime about climate variability in an easy-to-read, big-picture fashion.

The global climate of the Earth's atmosphere is extraordinarily complex and changeable. The sun's radiation is the primary heat energy source for our planet. Temperatures vary as a result of fluid dynamics—the turbulent circulation and inertia of air and sea—set in motion by the Earth's rotation, tilt, wobble, orbit, and orbital eccentricity. Then add to this equation the complex interaction of these air and sea fluids with land, above and below the sea.

Cloud physics is another of the tough areas. To date, computer simulations of global climate don't properly account for water in the air. Even more perplexing is the evaluation that "climate models have not improved in 50 years."[1] Simply stated, climate models don't match present or historical evidence. As a result, they don't properly predict the future, either.

One of my big worries is miscommunication. In our information age, people are more able to see what's happening around them on a grand scale. Thanks to computers, Internet technology, and the news, we see all the wild weather that's happening around us locally, regionally, and worldwide almost instantaneously. You name it—hotter temperatures, colder temperatures, droughts, floods, wildfires, hurricanes, tornadoes, etc.—we see them all, as they happen, up close. The reality of such events hasn't changed. They've always been there. What has changed is our awareness of the events. Such newfound awareness is worthwhile from the perspective of better warning and life saving, injury prevention, and property loss measures. But, a downside is that now we are apt to take a much narrower, tunnel-vision view of the present **weather** and not bother to look at **climate** history. For example, as we are bombarded with news and views that define weather outside our windows, we tend to quickly conclude that these events are worse than in the past and that they are even catastrophic and human-caused. But are they?

Another point that bothers me is that some world leaders and organizations seem to think they need to tax carbon—they really mean carbon dioxide (CO_2)—before "it's too late." They cite humans, and the CO_2 humankind generates, as the culprits for more intense hurricanes, tornadoes, wildfires, the price of rice, and on and on. Yet, what we do know about the atmosphere, from real scientific evidence, both current and historical, clearly shows that **CO_2 is NOT a problem.** Rather, CO_2, as well as water and warmth, all are indispensable to nourish life.

Weather vs Climate

I've also been frustrated with the improper usage of the terms weather and climate. Many use the terms interchangeably, which causes unnecessary confusion and angst. They are not synonymous terms. Let's straighten that out right now.

Weather, such as temperature in your own backyard, is a **daily, weekly, or even yearly event.**

Climate is a **long time and large area** average of daily weather events.

Said in another way:

Temperature is a specific atmospheric variable or event.

Weather is temperature, precipitation, wind speed, etc., at a specific location and time. It's what we see, feel, and react to each day.

Climate is weather averaged over a long period of time and a particular surface region of the Earth. It is a necessary metric for a more complete and systematic understanding of what is happening around us, and its variability. It's the memory of what we have seen, felt, and reacted to, amassed over many years.

Keep the fundamental difference between **weather** and **climate** handy. It will make understanding what follows much easier. Throughout this book, the terms climate, global warming, and global cooling will be used to mean temperature averaged over the entire surface area of the Earth, and averaged over at least a 30-year time period (the longer the better).

Climate Terminology

Similarly, the terms global warming, climate change, and climate variability will be used in order to present various sides of the story.

There are subtle, but important, differences among these three additional terms to be aware of as you read.

Global warming is a term that was used extensively since the late 1980s through 2009 or so. The original intent was to describe only the warming side of the cyclic temperature phenomenon. Global warming is an incomplete expression. However, the term still is used, as appropriate, to describe that part of the story. But it's only half the story. Global cooling is the other half of the cyclical behavior of our atmosphere.

Climate change is now in vogue to replace global warming. But these two terms dance back and forth in use. The term "climate change" is an attempt to consider both warming and cooling. However, the word "change" has a distinctly negative connotation to it. "Change" suggests something that is different, something that hasn't happened before. But, **Earth's climate has been changing for 4.6 billion years.**

Climate variability is the term I prefer. It is the best term I am aware of, as I believe the word "variability" has no negative connotation associated with it. It connotes that temperature varies in both directions, for whatever reason, and always has.

Be aware of the differences among the terms.

Lastly, keep in mind the acronym CAGW, another incomplete and manipulative term. It doesn't seem to be as popular right now but it does describe one extreme side of the story:

C: catastrophic (disastrous, awful, life-threatening, or ruinous)
A: anthropogenic (human-caused or not natural)
G: global (as opposed to regional or local climate)
W: warming (one half of the cyclic climate picture)

While the details of weather and climate may be complex, climate trends are easier to grasp when shown over many decades, many centuries, many millions and even billions of years, as you will see throughout this book.

My hope is that everyone will gain and enjoy a greater understanding of our incredible Earth-Atmosphere System and its miraculous ability to sustain and nourish all life on our amazing planet.

chapter 1

A Lifelong Interest

Ever since I can remember, I've been intrigued by our atmosphere and its weather. As a small child, I feared lightning, often hiding my head under the covers during late-night storms in order to muffle the sound of thunder. Thunderstorms are frequent spring and summer events in the Midwest where I grew up. But over the years fear gradually transitioned to excitement, thanks to my parents' gentle and calm portrayal of storms as fun to watch while safely indoors. They were simply typical weather phenomena, and we dealt with it. As a youth, thunderstorms became great fun, a thing to behold and watch in awe, like 4th of July fireworks. I was actually sad when storms, like fireworks, ended.

I've lived a lifetime of observing and researching the properties of the atmosphere, and I've been amazed by Mother Nature's wonders ever since I watched lightning and thunderstorms as a kid.

Growing up a mere three city blocks from the end of the main runway at Chicago's Midway airport, which happened to be the world's busiest airport back in the 1950s, was entertaining, too. In those days, one to two departing or landing planes would pass overhead every minute, at what seemed like barely above treetop level to a 10-year-old. I never noticed, or minded, the noise. Rather, I marveled at how the air I breathed could possibly support those monstrous flying machines.

As a teenager, I even learned to fly an airplane. By then, the newer and much larger O'Hare International airport, at the opposite end of the town, had taken over much of Midway's commercial air traffic. Midway, down to only about two to three commercial flight arrivals and departures per hour, was a fairly quiet general aviation airport to grow my wings. Quiet or not, anyone who has ever landed a tiny, 2-seat, Cessna 150 aircraft among commercial jet traffic knows about the bumpy air we call "wake" turbulence. That's turbulent air spiraling behind the wingtips of the bigger jets. There also is turbulence generated by the heat island effect of an airport located in the middle of the huge concrete and asphalt jungle of metropolitan Chicago. The two make taking off and landing a distinct challenge. Mastering the skill of flying in clear, but turbulent, air became yet another atmospheric "thrill" for this aspiring pilot and scientist. But like thousands of other private pilots flying in and out of Midway airport in small aircraft, the physics of clear air turbulence was a reality, and I adapted to it.

By 1967, I was a junior in the physics program at DePaul University and preparing for graduate school in meteorology at Florida State University after finishing my Bachelor of Science degree. My enchantment with weather did not falter or fade.

And More Weather Came My Way

The Chicagoland area where I grew up was at the northern edge of the region called "tornado alley." The ferocity of a tornado, dipping down from the clouds with its violent spinning motion and destroying anything and everything in its path, always intrigued me. I routinely observed many a tornado watch and a handful of tornado warnings back in the day. But, routine turned to soul-wrenching agony on a peaceful spring day in April 1967, when a devastating tornado stormed into my life, killing 33 innocent men, women, and children. (There is much more about this specific event, and tornadoes in general, in Chapter 10.)

"Stormfury" was my next adventure with turbulent weather phenomena. Flying into the eye of a 140-knot hurricane at a mere 1,000 feet above the ocean's churning surface is not a normal thing to do. Fortunately, I was on a Lockheed Super-constellation with some incredible Navy pilots who could safely "feel" their way in and out of hurricanes.

While a meteorology graduate student at Florida State University, I also interned with the U.S. Navy Weather Research Facility at Norfolk, Virginia. They, and their big brother, the National Hurricane Center, selected me as one member of a team of meteorologists for a hurricane weather modification project aptly nicknamed "Project Stormfury." We flew as crewmembers with the Navy's Weather Reconnaissance Squadron Four, the famous Navy "Hurricane Hunters," to RADAR-monitor and record the seeding of 1969's hurricane Debbie. What an unbelievable stroke of good fortune—flying and performing hurricane modification experiments to try and "steer and soften" a hurricane's intensity. I once thought turbulence was exciting during landings at Midway airport in Chicago. Hah! (There is a lot more about hurricanes in Chapter 11.)

After one more school year at Florida State, the U. S. Army called me to active duty. I ended up staying for 22 years as a career Army officer and soldier-scientist, and another five years as an Army civil servant. Those years included a variety of atmospheric-related, jobs such as developing radioactive fallout prediction tools and chemical agent diffusion research, computer simulation, and testing at Deseret Test Center; teaching and evaluating smoke, flame, and nuclear operations with the 25th Light Infantry Division; developing and testing nuclear weapons and analyzing their effects at Lawrence

Livermore National Laboratory and the Nevada Test Site; and the research, development, and production of military lasers and optics for ranging, targeting, and countermeasures at the Army's Program Executive Office for Aviation.

Also sandwiched in the midst of that first career was an Army-sponsored Master of Science degree in physics from Georgia Tech, as well as a visiting scholar appointment at Stanford University's Ginzton Laboratory of Applied Physics. Not surprisingly, the common thread through all those positions was the scientific make-up of the atmosphere.

Then it was on to a second, nearly 20-year career in the aerospace industry that involved more applied physics. This time, my focus had an atmospheric as well as outer space thread to it. It now included the development and production of laser offensive and defensive devices for military helicopters and fixed-wing aircraft, as well as system, subsystem, safety, security, and failsafe work for the nation's fleet of Intercontinental Ballistic Missiles (ICBMs).

Counting the two careers and the last four years of retirement that has focused almost entirely on my own independent research and the writing phases of this book, I've been in the atmospheric physics business for just north of 50 years total, and I've learned a lot.

Hence, I've got a lot I want to share.

Some of What I've Learned

Scientists seem to have a difficult time communicating science, especially to the general public, whether young or old. For years, I've thought that all scientists needed to learn to communicate better. Albert Einstein once said: "If you can't explain it simply, you don't understand it well enough."[1]

This is where my wife Nancy comes into the picture. Of this writing duo, I'm the scientific jargon wordy guy, and she is the English scholar who might as well have a PhD in clarity and common sense. I write; she edits; then I rewrite as many times as it takes until she likes it (smiles). The result is this book about climate science that's easy to read, and which plugs a gaping hole on the bookshelf.

These days, it seems that everything is blamed on global warming. For instance, when you hear the term "climate change," do you think of "global warming" and fear the Earth is becoming so hot we can't possibly survive? After all, aren't we killing off the polar bears because the ice caps are melting? Well, that's old news and wrong besides. The polar bears are flourishing. Their numbers are "the highest they've been in decades."[2]

There is good scientific research information and history that has been tamped down, misrepresented, and even lost in the noise of the politically driven "global warming" or "climate change" or "climate catastrophe" confusion.

Confusion ensues when, for example, some declare that a rise in the level of CO_2 is followed by a rise in temperature. Increasing CO_2 does not cause increasing temperature. It's the other way around. Temperature controls CO_2. A change in temperature (the cause) results in a change in CO_2 (the effect). Think of a bottle or can of soda and what happens to the drink's carbonation (CO_2) when you leave it in the sun. The soda warms and the carbonation departs the drink, and quite rapidly, I might add.

The big boys—sun, water, gravity, and inertia—control our temperature. CO_2 is a greenhouse gas that plays a negligible role in comparison to the big boys. I will have much more on this whole topic in Chapters 6, 8, and 9.

If you are over the age of 60, you have most likely experienced modest and short periods of global warming, modest and short periods of global cooling, and modest and short periods of relatively flat, neither warming nor cooling, global temperatures. We've had several instances of this type of cyclic behavior over the last 62 years. That's what our short-term weather and our long-term climate does, and has always done, for the last 4.6 billion years. It changes in a cyclic fashion.

Here are a few examples of the increasing (➚), decreasing (➘), and flat (–) global temperature trends over short periods during the last 62 years.[3, 4]

Year	Trend	Notes
1957-1964	–	
1964-1969	➚	
1969-1976	➘	The last global cooling scare (Front page of Time Magazine—"The Ice Man Cometh.")
1976-1979	➚	
1979-1985	➘	
1985-1987	➚	
1987-1997	–	Beginning of the current global warming scare
1997-1998	➚	
1998-2012	–	The infamous "pause" in the global warming scare (The switch from "global warming" to "climate change" terminology began in the last few years of this timeframe.)
2012-2016	➚	
2016-2017	–	
2017-2019	➘	

Figure 1-1. Global temperature trends over the last 62 years.

People worry because they are inundated with unsubstantiated claims that our global temperature is warming catastrophically. The truth is that we are warming, but just slightly, normally, and beneficially over the long haul.

The information provided in this book challenges catastrophic global warming with clear, precise, and verified scientific evidence of a natural, normal climate. The discussion includes information about the air we breathe and the role of the sun, water, gravity, and inertia.

Also included in the discussion are the greenhouse effect, the part that carbon dioxide and water vapor play in the greenhouse effect, climate data manipulation, climate-related policy, the most recent developments that surround these issues, and much more.

The topic will be explored using common sense, a big picture perspective, and the time-honored **scientific method**. Facts will be sorted out from fiction using real observations. The scientific method is as important today as ever and the only way to truly understand our planet's climate. It is the accurate way to capture the ice-cool confirmation of real science and dump the volcanic ash of agenda-driven pseudoscience.

What You Will Find

This book compares and contrasts many of the diverse angles of this complex and controversial topic. To do that, climate reality will be explained with thorough straightforward language and balanced with graphics that illustrate eons-old cyclic warming and cooling behaviors of the Earth's atmosphere. The divisive subject of global warming will be demystified through a series of simple, scientific discussions with conclusions illustrated with graphs, definitions, pictures, cartoons, and comparisons. A well-defined picture will emerge showing what we do, and don't, know about our ever-changing **global climate**.

We will cover the full spectrum of global climate science and associated current events. As you read, please keep this very important point in mind: This book focuses on global climate variability, not local or regional **weather** variability. Of course, I'll discuss weather, too. But, as stated in the Author's Note and worth saying again, weather and climate are not the same. Both weather and climate will be discussed so that you can see the differences. However, always remember that they are not interchangeable terms.

A simple, sound science, common sense perspective will provide you with lots of relevant, fun, and useful information. Our book will give you a lot to think about.

chapter 2

The Scorecard

Acronyms

There are many acronyms in this book. I've listed all of the ones used in each chapter at the start of each chapter. Here are the ones used in this chapter:

As they say in baseball, "You can't tell the players without a scorecard."[1] This chapter will present the global climate scorecard, up front, so that the rest of the book can be as clear, concise, easy-to-read, and as useful a book as possible. This scorecard begins with three sciences that apply to this book, followed by definitions and other clarifications that are critically important to the topics. We don't want you scratching your head, searching for, and identifying pieces of the puzzle and then wondering how the puzzle pieces all fit together. I do this upfront as a foundation. Then you know where to find things easily.

ACRONYMS USED IN THIS CHAPTER

Acronym	Definition
CO	Carbon Monoxide (Gas Molecule)
CH_4	Methane (Natural Gas Molecule)
CO_2	Carbon Dioxide (Gas Molecule)
°F	Degrees Fahrenheit
NOx	Family of Nitrogen Oxide Gas Molecules
O_2	Oxygen (Gas Molecule)

Acronym	Definition
O_3	Ozone (Gas Molecule)
P	Pressure
SOx	Family of Sulfur Oxide Gas Molecules
T	Temperature
UAH	University of Alabama (at) Huntsville
V	Volume

Fields of Applicable Science

One of the truths of science is that it can be quite complicated. For example, mixing the three fields of natural science together—physics, chemistry, and biology—is complicated. Adding multiple science specialties, such as sun science, Earth science, and atmosphere science, drives the complexity factor up remarkably fast. In this particular example, there are nine combinations of fields and sub-specialties. Those are three fields of natural science multiplied by the three sub-specialty areas that apply to each field. And, any one of these nine combinations could easily occupy the entire career of a scientific specialist, such as a solar physicist.

Definitions

Let's start with definitions, first of the three natural science fields.

Physics is the study of matter and its motion in space and time.

Chemistry is the study of the composition, structures, and properties, of matter.

Biology is the study of life and living organisms.

The following is a list of definitions that can pertain to any of the hundreds of science sub-specialties in this book. These definitions also fit within the realm of each of the natural sciences.

Anthropogenic: An adjective used to describe something that results from the influence of human beings, rather than nature. Truck exhaust is a good example of an anthropogenic pollutant.

Atmospheric Physics: The physics of the atmosphere (aka physical meteorology). This is an example of a sub-specialty of physics.

Atmospheric Pollutants: A cocktail of mostly visible gas, liquid, and particulate chemicals in the lower atmosphere that cause harm or discomfort. Some manmade and some are produced naturally. The gas and liquid forms of pollution include an assortment of nitrogen and sulfur oxides (NOx and SOx), elemental carbon and carbon monoxide (CO) from incompletely burned fossil fuels, and the ubiquitous ozone molecule (O_3). The airborne particulate matter included in this definition includes dust, elemental carbon (soot), and smoke from wildfires, etc.

- **Smog:** The gas and liquid subset of mostly visible atmospheric pollutants.
- **Haze:** The particulate subset of visible atmospheric pollutants.

Climate: The prevailing weather conditions, averaged over a long period of time and a large area of the Earth's surface.

Climate Model: A mathematical simulation of climate.

Evidence: Actual observations or measurements of atmospheric parameters such as temperature, humidity, and carbon dioxide concentration.

Fossil Fuels: Natural solid, liquid, or gas fuels, such as coal, oil, and natural gas that are formed in the earth from dead plants or animals.

Fossil Fuel Combustion: Is the process of igniting fossil fuels, such as coal, oil, or natural gas, in the presence of oxygen to produce heat energy for the home, workplace, and vehicles.

Figure 2–1 is the chemical formula for the simplest of the fossil fuels—the complete combustion of the natural gas, methane, and oxygen (O_2) to produce heat energy, carbon dioxide (CO_2), and water vapor (H_2O). None of these three products are pollutants. Instead, all three of these fossil fuel combustion products are necessary to sustain and nourish all life on our planet.

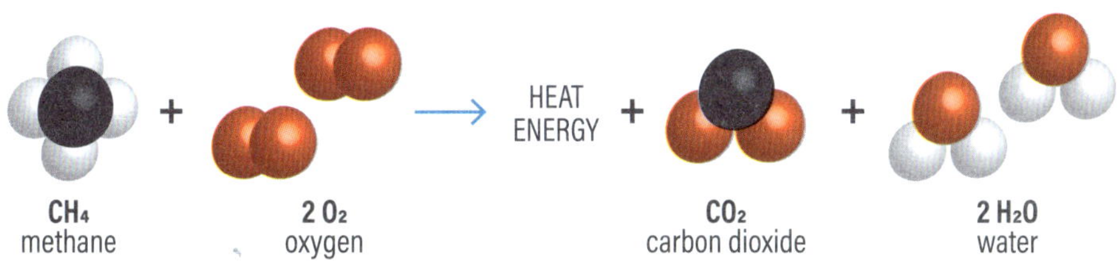

Figure 2-1. The chemical formula for the complete combustion of natural gas.

The more complex fossil fuels (e.g., gasoline, oil, and coal) also generate heat, CO_2, and H_2O. However, these complex fossil fuels also generate a variety of atmospheric pollutants as additional byproducts of combustion. These pollutant byproducts include partially burnt fuel elements, namely elemental carbon (aka soot) and carbon monoxide (CO), as well as smoke, sulfur and nitrogen oxides, and volatile organic compounds (VOCs). **Note that CO_2 often is incorrectly grouped with these pollutants, even though it is an invisible, essential, non-toxic, and non-polluting gas of life, just like H_2O.** There will be much more on pollutants and non-pollutants and how they do and do not affect climate in Chapters 7 and 8.

Global: The entire surface area of the Earth, including both land and water.

Geology: The study of the Earth, its structure and its life.

Geophysics: The physics of the Earth. This is another example of a sub-specialty of physics.

Graph: A plot of large amounts of data in a concise manner.

Gravity: The attractive force between two or more bodies.

Ice Ages: Alternating long, cold, glacial periods and shorter, interglacial warm periods:

- **Glacial periods:** A cyclic, long period of time characterized by cooler temperatures (typically about 10–12 °F below normal) in the lower atmosphere. Advancing snow and glacial fields distinguish an ice age, which typically lasts 70,000 to 100,000 years, followed by a 12,000–18,000 years-long interglacial warming period before the next ice age.
- **Interglacial periods:** A cyclic, long period of time characterized by warmer temperatures (typically about 10–12 °F above the depths of ice age temperatures) in the lower atmosphere. Retreating snow and ice fields distinguish an interglacial warming period, which typically lasts 12,000 to 18,000 years, followed by another 70,000 to 100,000 years-long glacial period before the next interglacial warming period. We are in the tail end of an interglacial warming period right now.

Ideal gas law: Describes the relationship among pressure (P), volume (V), and temperature (T) of the ideal gases that make up our air.

Meteorology: The study of atmospheric processes and phenomena.

Oceanography: The study of the physical and biological aspects of the ocean. It includes the depth and extent of ocean waters, their movement and chemical makeup, and the topography and composition of the ocean floors.

Orbital motion: The motion of planets, and specifically the Earth, around the sun.

Paleoclimatology: The study of ancient climates.

Paleontology: The study of ancient forms of life.

Photosynthesis: The physical process of synthesizing oxygen and food from vegetation, sunlight, water, and carbon dioxide.

Simulation: A theoretical representation or model of a system, such as weather or climate, that allows for investigation and analysis of the properties of the system and, in some cases, prediction of future outcomes.

Sub-specialty: A specific component or specialty of one of the three natural science fields of physics, chemistry, and biology.

Tectonics: Forces within the Earth that cause movement or deformation of the Earth's crust, including our sea floors.

Thermodynamics: The sub-specialty of physics dealing with heat and temperature.

Types of scientific statements:

- **Hypothesis:** An uncertain explanation of an observed phenomenon for further investigation.
- **Theory:** A demonstrated explanation of a set of hypotheses.
- **Law:** A demonstrated explanation of a natural occurrence that is universally true.

Vegetation: Refers to all forms of plant life, their growth, and ultimate decomposition.

Weather: The "state" or condition of the atmosphere at a specific location and instant of time. It's what you see and feel, either looking out the window or being outside on any given day.

As you can see, there are a multitude of interconnected disciplines that collectively and systematically contribute to the Earth and its atmosphere, weather, and climate complexity. What we have are diverse, yet interconnected, areas of science that are not always well understood either individually or collectively.

The fact that we humans don't have perfect knowledge doesn't help either. Albert Einstein said it best: "We still do not know one thousandth of one percent of what nature has revealed to us."[2]

But that doesn't mean that it is hopeless to read any further. It just means the reveal has just begun. We need to look closely as best we can at what the Earth's actual climate behavior is really trying to tell us. My job is to sort the puzzle out piece by piece in easy-to-understand bite-size pieces of words and pictures, and then reassemble them in an equally unadorned manner.

Clarifications of Important Definitions

Graphs

Graphs are pictures used to efficiently portray large amounts of data (numbers) in a simple, compact, easy-to-understand manner. For example, sitting in front of me as I write this chapter are nine pages

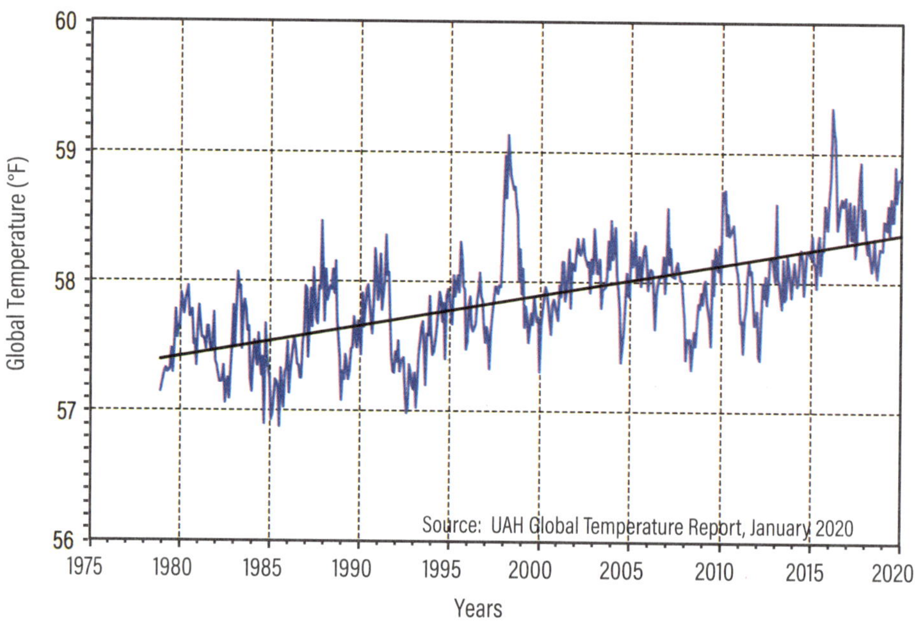

Figure 2-2. Satellite-based global temperature for the period December 1978 through December 2019.

of paper that contain 480 rows of numbers representing month, year, and global temperatures that have been measured by radiometers on weather satellites for 40 years. With a computer, it took all of five minutes to convert the electronic version of those nine pages of original temperature data to the graph in Figure 2–2.[3]

This is a 2-dimensional plot of satellite-based average global temperature during the 41-year period of temperature measurement from December 1978 through December 2019. The horizontal, or "x", axis of this plot is time in 5-year increments. The vertical, or "y", axis is the lower troposphere temperature scale in 1-degree Fahrenheit (°F) increments. The blue zigzag curve and accompanying black straight line illustrate temperature variation and trend, respectively, over the period. Note the fairly wide variation in global temperature over the period (roughly ±¾ °F) and the gently rising temperature trend of approximately 0.04 °F per year, to a value of nearly 59 °F by the end of 2019.

Weather vs. Climate

Weather and climate have many similarities, but also a few distinct and very important differences, as highlighted in light blue in Figure 2–3.

Weather and climate are the same in terms of state (condition) of the atmosphere and geographical coverage, but you can get in trouble really fast by ignoring the time period differences and mixing weather and climate talk.

Weather is what we see on the nightly news (e.g., typically hourly, daily, and weekly reports and forecasts on temperature, pressure, humidity, rain, snow, wind, severe storms, and the like). Climate, on the other hand, is usually about historical record trends regarding the data sources that, as far as we know, go back over hundreds, thousands, and in some cases, even millions to billions of years.

Characteristic	Weather	Climate
Variables	Temperature	
	Pressure	
	Precipitation	
	Wind speed & direction	
	Cloud cover	
Time Period	Short-term	Long-term
	Instantaneous	Average
	Hourly, daily, weekly, monthly, even yearly	Multiple decades (at least 30 years)
	Can change rapidly	Changes more slowly
Geographical Coverage	Local	
	Region	
	Country	
	Continent	
	Globe	

Figure 2-3. Comparison of the terms weather and climate.

As a result, we have enough empirical data to paint a fairly decent picture of historical climates. However, forecasting future climate, which is only accomplished via computer simulations, is fraught with peril. There is simply too much atmospheric physics that we do not yet fully understand. Therefore, my advice is to be very wary of those who predict future climate that is based on the current maturity of computer simulations, especially if the simulations are not able to successfully model (aka hind cast) known climate history.

Pollutants vs. Carbon Dioxide

It is a quite common mistake to equate ugly and toxic air pollutants such as dust, smoke, soot, sulfur dioxide, and so on, with carbon dioxide, which is an invisible, non-toxic, clean, and useful molecule. We drink it in carbonated beverages. We generate it with every breath. Plants convert it to oxygen. Carbon dioxide is necessary for all life on this planet.

Global Warming vs. Climate Change vs. Climate Variability

Global warming is the battle cry of those who perceive a dangerous upward trend in surface temperature, be it local, regional or global in nature. It was an early-on (1990s), oft-repeated slogan used to hype computer models that said the Earth was warming at an uncontrolled, runaway, even catastrophic rate, as a result of humans' unbridled use of fossil fuels. But then the "pause" arrived. Measured global temperatures—the physical evidence—didn't significantly rise or fall for twelve years. Instead, they flat-lined, or paused, from their increasing trend between 1998 and 2010, in spite of a continuing rise in atmospheric CO_2 levels. Furthermore, history is replete with both warming and cooling episodes such as day and night, summer and winter, El Nino and La Nina, and ice ages and interglacial warming periods, to name a few. And the climate models did not accurately portray them. So, those touting global warming switched to "climate change" about ten years ago. **Climate change** supposedly allows for temperatures to warm, cool, or stay the same. It is better in that sense.

But, "change" is still a word loaded with ideological baggage and a one-way (negative) connotation. It suggests that something different, and more often than not something bad (e.g., an out-of-control climate), is happening and humans are to blame. I prefer the phrase **climate variability** as a more natural expression that accommodates both ups and downs. Consequently, I will use the term "climate variability" wherever appropriate in this book as a more accurate representation of global temperature measured over climatic time scales. When you see global warming or climate change mentioned from this point forward, it is a reference to someone else's wording. Our climate is now, and always has been, variable in a controlled, cyclic fashion.

chapter 3

How Scientists Determine Truth

The Scientific Method

"Science without doubt isn't science at all . . . The scientific method is the way we tackle doubt in order to reduce it."

—Mike Rowe[1]

Opinions may be brilliant, but most are like eyeballs, that is, everybody has them. But whether or not you can see straight with them is another story all together. That's why we scientists have a process called the scientific method to establish truth, rather than speculate, about the world around us.

"The scientific method is a simple process that has been used for centuries. It involves identifying a question, developing a hypothesis, constructing an experiment to confirm the hypothesis, and analyzing the results. If the results do not align with the original hypothesis, the hypothesis must be re-examined . . ."[2]

The scientific method welcomes critiques so hypotheses can be refined. And it avoids speculation about distant events for which there is no hard evidence.

ACRONYMS USED IN THIS CHAPTER

Acronym	Definition
AMS	American Meteorological Society
CAGW	Catastrophic Anthropogenic Global Warming
NASA	National Aeronautics & Space Administration

Before moving forward, I also think it is useful to clarify the often misunderstood differences between scientific hypotheses, scientific theories, and scientific laws. Figure 3–1[3] lists short definitions of each, and several other criteria that support the definitions.

	MAIN TYPES OF SCIENTIFIC PROCLAMATIONS		
	Hypothesis	**Physical Theory**	**Physical Law**
Definition	Uncertain explanation of an observed phenomenon for further investigation.	Demonstrated explanation of a hypothesis, including how and why it is true.	Demonstrated statement of a natural occurrence that is universally true.
Examples	Earth's climate is warming.	The special theory of relativity.	The photelectric effect.
Purpose	Present a possibility that can be explored.	Describe a large set of observations.	Describe a large set of observations that are consistently made.
Basis	Research.	Research and reliable, reproducible, and independent testing.	Extensive research, and reliable, reproducible, independent testing.
Scope	Single, specific event.	Various, specific events.	General events.
Testable	Yes	Yes	Yes
Falsifiable	Yes	Yes	Yes
Well-tested	No	Yes	Yes
Data	Limited	Wide set of experimental data, tested under various circumstances.	Wide set of experimental data, tested under all known circumstances.

Figure 3-1. Types of scientific proclamations.

All three are similar, except for scope and amount of testing. A hypothesis also differentiates itself from the other two by virtue of the fact that a hypothesis essentially begins the scientific method.

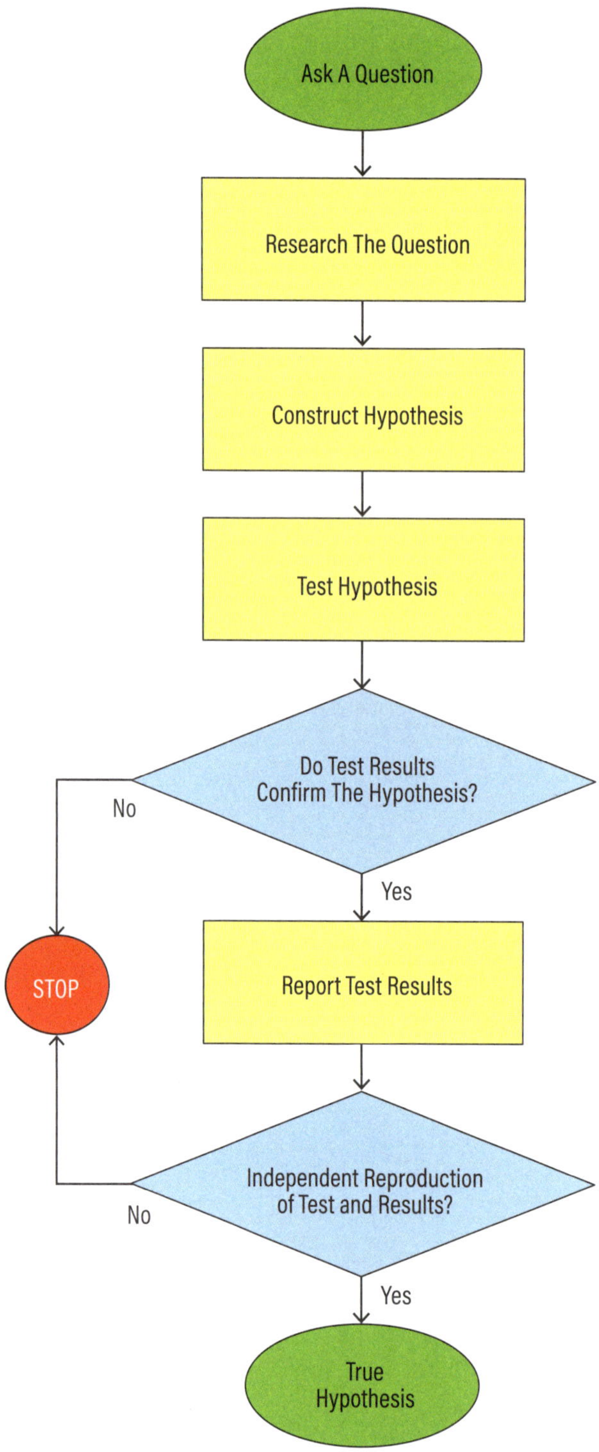

Figure 3-2. The "Scientific Method."

The other two, physical theories and physical laws, are results of a successful application of the scientific method.

Figure 3–2 is a flowchart that represents the scientific method. It is the process that asks and answers a hypothetical question, and confirms both theories and laws.

First, you pose a question about something that you do not understand. Then you study, or research, the issue to find out what others may think about it. Based on that, the next step is to formulate a hypothesis, or statement, of your own about what you now believe is the truth about the topic. The hypothesis should be a succinct and specific statement of what you believe to be true.

In 1887 German physicist Heinrich Hertz hypothesized and demonstrated the "photoelectric effect"—the transfer of energy from light to electrons in a material that generates electricity. The photoelectric effect became a law at that time because it worked universally, even though Hertz did not understand why that transfer of energy from light to electricity happened. It wasn't until 1905 that another German physicist, Albert Einstein, won a Nobel Prize for his explanation of why the photoelectric effect converted light to electricity. At that time, the photoelectric effect also became a demonstrated physical theory. Today we now have photocells, solar panels, night vision devices, and more, as a result of the use of the scientific method by Hertz and Einstein.

Next, you develop and conduct an experimental test. This step provides physical evidence that will either accept or reject the truth of a hypothesis. Evidence includes observations of things seen, held, heard, and recorded by people at specific times and places. This is unlike "models" that are all too often nothing more than computerized hypotheses that are never validated, i.e., tested, to ensure they reliably reproduce reality. Experimental testing is an absolutely necessary step in the scientific method.

After completing the test, the results are analyzed. From the analysis, a conclusion is drawn as to whether the test confirms or denies your hypothesis. At this juncture, if the test demonstrates the hypothesis to be false, it is time to stop, or start over, with a revised hypothesis. If the test demonstrates the hypothesis to be true, this is good news. But, this is not the end of the scientific method.

Results are now reported to the scientific community. This is so independent scientists can reproduce the experimental testing of the hypothesis to verify and validate the results of the test. The details of how the test was set up and executed must be disclosed, along with all the resultant test data. At this juncture, if the results are reproduced, the truth of the hypothesis has been successfully defended and is deemed to be correct.

The last point about the scientific method is that it is never-ending. If someone comes along years, decades, or even centuries, later and successfully negates the hypothesis, then the scientific method must be revisited. That's how science works, and has worked since the early 16th century when Francis Bacon introduced the scientific method. It's never done. It's never over. Science is never "settled."

An example of this is what Albert Einstein said about the book *One Hundred Authors Against Einstein.* He said: "Why one hundred? If I were wrong, one would be enough."[4, 5, 6] His statement pertained to his Special Theory of Relativity, and to this day, Einstein's theory has never been demonstrated to be wrong. But that doesn't mean that it won't happen tomorrow.

Yet, many now ignore this basic tenet of science in favor of consensus or popular opinion in order to justify a claim or promote a personal agenda, even if evidence doesn't support it.

No Consensus

The Merriam-Webster Dictionary defines consensus as: "A general agreement about something; an idea or opinion that is shared by all the people in a group."[7] Today many people accept consensus as "social proof" for almost anything. However, while somewhere there may be a role for consensus, and its partners "popular opinion" and "groupthink," there is no role for it in real physical science and the time-honored scientific method.

Recall the scientific skeptic, Galileo. The popular opinion of his time was that the sun revolved around the Earth. He said, no, the Earth orbited around the sun, and his newly developed telescope demonstrated it. Nevertheless, he was thrown in jail for a time for sharing his belief. The crime was heresy. Still, Galileo continued to study, learn, experiment, and write secretly. In the centuries since, science has confirmed that he was right. The Earth does orbit the sun in a slightly elliptical manner, which has a big effect on our climate. There will be more details on planetary motion in Chapter 5.]

Being quantitative and skeptical is the lifeblood of real science.

More recently, the Challenger space shuttle accident occurred on the unusually chilly Florida morning of January 28, 1986. It exploded a mere 73 seconds into its flight, killing all seven astronauts on board.

In the investigation that followed, investigators determined that "groupthink" was the weak link that triggered the decision to launch. The investigators found that the O-rings in the shuttle/booster motor assembly malfunctioned due to the uncommonly cold environment that morning. The launch could have been delayed to wait for a warmer time window.

The 97% Myth

The application of "truth by consensus" to physical science has seemed to mushroom in popularity only since the latter years of the 20th century. And it was probably as a result of deep, albeit natural and healthy, disagreements among scientists on multitudes of modern topics. But the problem is that consensus does not establish truth, scientifically. That's because science requires the use of the scientific method to ascertain truth.

So, let's talk scientifically about a current example of consensus gone awry—Catastrophic Anthropogenic Global Warming (CAGW). The hypothesis is that our planet is warming in an uncontrolled and unbounded manner, and human use of fossil fuels is the cause. Red flags erupt right off the bat. This hypothesis is too vague. And testing of the hypothesis is extremely difficult, if not impossible. How much warming is considered catastrophic is not defined or agreed upon, nor is how to measure it. Furthermore, there is no known way to fingerprint the natural from a human role in the warming process. But, the real elephant in this hypothesis room is the time period of warming. Is it 10 years, 100 years, 1,000 years, or maybe more? This unanswered question, alone, makes a proper experiment impossible.

A hypothesis must be clearly and unambiguously stated at the start in order to achieve truth at the end. Otherwise, insufficient or flawed testing, sloppy analysis, and ultimately wrong conclusions result. Add to that a poorly constructed, executed, and analyzed survey, and what you get is the misleading urban myth that 97% of all scientists believe that humans are changing the Earth's climate and fossil fuels are the cause.

Below is a blow-by-blow look at the surveys that show their claim of 97% consensus is nothing short of pure exaggeration and spin for the usual pursuit of money, power, control, and careerism.

The first 97% consensus survey was a non-peer-reviewed, opinion essay written by the historian Naomi Oreskes that appeared in the journal *Science* in June of 2004.[8]

Seventy-five of 1,117 abstracts that she chose to examine either **explicitly** or **implicitly** supported human-induced climate change over the last 50 years. Her search criterion was "global climate change" and was silent about whether climate change was dangerous, benign, or good. Notably, she did not examine published papers, rather only abstracts, which are notorious for overstating research findings to gain notice. Furthermore, the review inexplicably missed hundreds of articles by prominent global warming skeptics.

Not surprisingly, a year later social-anthropologist Dr. Benny Peiser reran Naomi Oreske's experiment. He found that 470 of the 1,117 abstracts were unrelated to climate change, except that, for whatever reason, they included the phrase "global climate change" in the abstracts. Another 634 abstracts were highly suspect, since they had only an implied agreement with human-induced climate change. The author of the survey, Naomi Oreskes is neither a scientist, nor a mind reader. Therefore, those 634 abstracts are not valid either. Dr. Peiser's conclusion was that only 13 of the 1,117 abstracts (1.2%, not 75%) in Oreske's badly flawed consensus survey **explicitly** agreed with human-induced climate change.[9, 10, 11]

Next came Peter Doran and Maggie Zimmerman. They conducted a short online survey in 2009 and presented their results in *EOS, Transactions, American Geophysical Union*.[12] They claimed 97.4% agreement with human-induced climate change. That percentage was based on responses by 75 of 77 self-identified, but not verified, **climatologists**. Doran and Zimmerman excluded 3,069 other relevant climate scientists, including meteorologists and physicists, that responded to the survey. This means that only 2.4% (75 of 3,146), not 97.4%, of respondents, agreed that global temperatures were rising and human-induced.

A year later, in 2010, William Anderegg, then a student at Stanford University, claimed that 97–98% of climate researchers publishing in the field supported human-induced climate change. He did so, along with three other co-authors, in a paper entitled "Expert Credibility In Climate Change" in the Proceedings of the National Academy of Sciences.[13] His was another Internet abstract counting drill. His survey of 1,375 abstracts claimed that 903 were convinced by evidence of human-caused climate change and 472 scientists were unconvinced by evidence. That's 66% agreement. However, he further filtered his results down to only the top 100 scientists with the most publication citations of which only three were unconvinced. Hence, his claim ultimately became another 97% proclamation.

Australian blogger and cognitive scientist John Cook, along with other contributors, performed yet another survey in 2013. The team examined the abstracts of 11,944 global warming/global climate change peer-reviewed papers that were published during the period 1998–2011. They subsequently reported in *Environmental Research Letters* that: "Among abstracts expressing a position on human-caused global warming, 97.1% endorsed the consensus position that humans are causing global warming."[14] The problem was that they excluded 7,930 of the surveyed abstracts from the percentage. They wrongly scored other abstracts, selectively sampled the population of pertinent abstracts, etc. That's no way to design, execute, score, or analyze a survey.

Subsequently, Dr. David R. Legates of the University of Delaware reexamined the Cook et al survey, and determined that only 41 of the 11,944 abstracts explicitly supported human-caused warming.[15] That's barely one third of one percent, or 0.3%, not 97.1%, consensus.

Multiple authors of inspected abstracts also issued blistering critiques of the Cook et al survey techniques. These authors disagreed with the Cook team's classification and scoring of their papers, since they did not endorse human-caused global warming.[16] And they said so in a variety of journals, media organizations, and weblogs.

The American Meteorological Society (AMS) weighed in with their own survey of member meteorologists and environmental professionals in 2014.[17] They reported that 52% of 1,854 respondents agreed with human-induced warming over the past 150 years. Strangely, this finding never made it into the published version of their peer-reviewed article.

Finally, George Mason University conducted a follow-up survey of AMS professionals in early 2016. They reported that 67% of 4,092 U.S. and international AMS respondents agreed that changes in the climate over the past 50 years are entirely, largely, or mostly due to human activity.[18] Still, survey consensus is not science.

Regardless of percentages, real science only takes one substantiated, reproducible set of evidence to bring down a whole consensual house of cards. Einstein's Theory of Relativity is a classic example of real science without consensus. No one has ever demonstrated Einstein being wrong.

Prophets of Doom

Why does this 97% meme persist? It's because fear mongering is a lucrative way to garner contracts, grants, etc. A common trick of those seeking money, power, control, or fame is to scare people into thinking the "sky is falling," or catastrophe is lurking around the next corner. Examples of this are the misleading assertions that storms such as tornadoes and hurricanes are becoming more frequent and intense. Author Note: These weather phenomena are **not** getting more frequent or intense, as I shall demonstrate, in detail, in Chapters 10 and 11.]

All too often, the 97% number is blindly cited to lend credibility to arguments, like severe storm numbers, without ever checking or confirming the authenticity of those claims. The result is an avalanche of **false "facts"** made even worse by misinformed advice given to world leaders who further spread the supposed consensus. Here are some examples:

2015: "Never have the stakes of an international meeting been so high, because it concerns the future of the planet, the future of life."

—**Former French President Francois Hollande** at a 150-nation climate summit in Le Bourget, France (Nov 2015).[19]

2015: "I am not sure, but I can say to you 'now or never'. Every year the problems are getting worse. We are at the limits. If I may use a strong word I would say that we are at the limits of suicide."

—**Pope Francis,** when asked if the U.N. climate summit in Paris (2015) would mark a turning point in the fight against global warming.[20]

2014: "We have 500 days—not a day more—to avoid a climate disaster."

—**M. Laurant Fabius,** French Minister of Foreign Affairs and International Development, in the National Assembly (May 2014).[21]

Whoops. Time's up!

2013: "So the question is not whether we need to act. The overwhelming judgment of science—of chemistry and physics and millions of measurements—has put all that to rest. Ninety-seven percent of scientists, including, by the way, some who originally disputed the data, have now put that to rest. They've acknowledged the planet is warming and human activity is contributing to it . . . Nobody has a monopoly on what is a very hard problem, but I don't have much patience for anyone who denies that this challenge is real. We don't have time for a meeting of the Flat Earth Society."

—**Former President Barack Obama,** at Georgetown University, June 25, 2013.[22]

Assertions aren't science.

2009: "Global warming will kill one billion people in a decade"

—**John Holdren,** testifying at his Senate Confirmation Hearing to become President Barack Obama's chief science advisor, February 14, 2009.[23]

The decade is now over.

2009: "The best projections tell us that we have less than 100 months to alter our behavior before we risk catastrophic climate change."

—**Prince Charles** of England in a speech to business leaders in Brazil.[24]

They just keep moving the goal posts.

1988: "In forty years (2028) the West Side Highway will be under water."

—**James Hansen,** NASA[25]

The water level of the Hudson River was ten feet below the highway in 1988. Sea levels are rising at a rate of 3 millimeters per year, or 0.01 foot per year. At that rate, the Hudson River will still be about 9.6 feet below the highway in 2028. See Chapter 12 for more on the rise of sea levels.

CHAPTER TAKEAWAYS

1. The scientific method has produced no scientific **evidence** of catastrophic, human-induced heating of the Earth's atmosphere. A consensus, without that cold, hard, reliable, reproducible evidence that comes from proper application of the scientific method, is worthless no matter the number or qualifications of the membership of the consensus.
2. The claim of a 97% consensus is a persistent and nonsensical myth that refuses to die in spite of its bogus/phony unscientific ascent.
3. Figure 3–3 summarizes the actual percentages of agreement with human-induced global warming.
4. Figure 3–3. Survey respondents/abstracts and percentage agreement with human-induced global warming.

Survey Originator	Year of Survey	Number of Respondents	Actual Agreement Percentage
Oreskes	2004	928	1%
Peiser	2005	1,117	1%
Doran & Zimmerman	2009	3,146	2%
Anderegg et al	2010	1,372	66%
Cook et al	2013	11,944	< 1%
American Meteorological Society	2014	1,854	52%
George Mason University	2016	4,092	67%

Figure 3–3. Survey respondents/abstracts and percentage agreement with human-induced global warming.

At different times, our climate warms and it cools. The sun, water, gravity, inertia, and orbital and tectonic motion, not humans, drive our climate in a cyclic and orderly fashion. Our climate responds to Mother Nature, not the social consensus of pseudo-scientists or some politicians and the "hot air" they produce.

Still, regardless of the source or magnitude of climate contributors, we humans must be good stewards of our Earth. By that I mean we all must advocate for clean air, land, and water. As such, I strongly advocate for the reduction of all types of pollutants. See Chapter 7 for details.

In conclusion, I think the words of this U.S. President summarizes this chapter much better than anything I could write:

> *"The great enemy of truth is very often not the lie—deliberate, contrived and dishonest—but the myth—persistent, persuasive and unrealistic. Too often we hold fast to the clichés of our forebears. We subject all facts to a prefabricated set of interpretations. We enjoy the comfort of opinion without the discomfort of thought."*
>
> **—John F. Kennedy**[26]

chapter 4

The Big Picture—Looking Beyond What We Can See

A Thought Exercise

Imagine, if you will, that you are a lone person on a remote tropical island. It doesn't matter how you got there, because you have permanent amnesia. While you don't remember a single thing, you do have food, water, a makeshift shelter, and fire for cooking and warmth. And no, you don't have any of today's electronic gadgets: no radio, cell phone, computer, Internet, Google maps, GPS, much less a telescope or binoculars. You know nothing of the past.

However, you are observant and inquisitive. You believe you are not moving and yet, each day you see an intense yellow object moving across the sky that provides light and warmth. You also note that it rises far off to one side of you, moves across the sky, and then settles far off to the other side.

What conclusion do you think you would draw concerning the ground you are standing on and the yellow object in the sky? You probably think that you and ground you stand on are not moving, and the yellow object in the sky (the sun), that is rising in the morning and sinking in the evening, is revolving around you and the land

UNITS OF MEASURE USED IN THIS CHAPTER

Acronym	Definition
T (°F)	Temperature (in units of degrees Fahrenheit)
$\bar{T}$	Temperature averaged over time and space [T(x,y,z,t)]

you are standing on. That is a reasonable conclusion. And it happens to be the exact same belief of **most** people on the planet, as late as the mid 17th century. This belief is called the geocentric, or Earth-centered, theory of planetary motion. I bolded the word "most" in the previous sentence for emphasis because while most believed the Earth was the center of the universe [Figure 4–1(a)], five very famous scientific doubters, skeptics, and deniers of the time did not believe it. Instead, they believed that the sun was the center of the universe [Figure 4–1(b)]. This was a heliocentric, or sun-centered, theory.

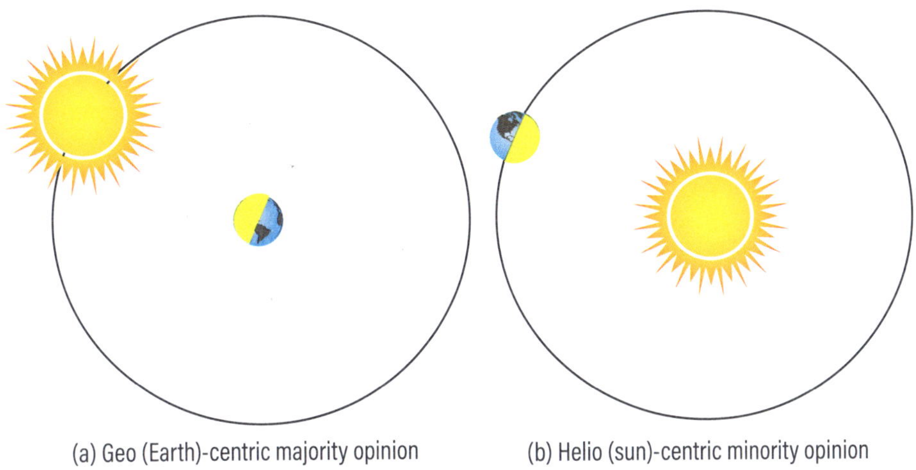

Figure 4-1. 17th century opposing views of the universe.

The skeptics were none other than scientists and mathematicians Nicolaus Copernicus of Poland, Johannes Kepler of Germany, Giordano Bruno and Galileo Galilei of Italy, and Sir Isaac Newton of England. Copernicus first hypothesized a heliocentric planetary system. Later, the other four, with the help of Galileo's invention of the telescope in 1610, were able to look far beyond their unaided eyes and prove Copernicus' theory that the sun was the center of the solar system, and that the planets, including Earth, orbited around the sun.

Early on, that was a highly controversial proclamation. The populous of the era burned Bruno at the stake for his heliocentric heresy. They jailed Galileo and forced him to renounce his newfound, sun-centered belief, lest he also suffer the same fate as Bruno.

But, what about the technologically advanced world of today, with all the cell phones, Internet, Google, radio, TV, and the like that are available? There should be no question that we live in a solar-based planetary system.

Yet as recently a 2018, a U.S. National Science Foundation poll[1] found that about 27% of the **U.S.** population still believed that the sun orbits around the Earth, rather than the Earth around the sun. And that's not all. The same poll determined that approximately 21% of the **world** population still believed that the sun orbited the Earth.

The point of this story is that the universe is a big place, and advancing technology has allowed us to look both microscopically small and telescopically large to determine truth. And in doing so, we're discovering that the volume above, below, and surrounding us is actually an interconnected and often complex system and not a bunch of isolated, disorderly events. Yet it is still relatively easy to be deceived, innocently or not. So, it is important to all of us to be mindful of the immediate, as well as the "bigger picture", in our lives.

Big Picture = x, y, z, t and T

Big picture means spatial coordinates "x," "y," and "z," and the temporal (time) coordinate "t." But, the big picture is much more than those four dimensions when discussing long-term climate. **Weather** is the "state" or condition of the atmosphere at a specific location and instant of time. Those conditions are what we now need to add to the discussion, namely heat (temperature), cloud cover, precipitation (rain, snow, etc.), wind (speed and direction), and pressure of the atmosphere. Now we'll take those daily, weekly, and even monthly local weather conditions and expand them to longer times and global space to get to climate.

Climate, then, would be the "average" of any or all of the above parameters over multiple decades of time. For example, local climate could be the local temperature or precipitation or any of the atmospheric parameters averaged over multiple decades of time. Similarly, global climate would be any of the atmospheric parameters averaged over both multiple decades of time (t) and all space (x, y, z).

A mathematician might characterize a global climate parameter, like temperature (T), as a function of x, y, z, and t. Said another way, temperature is a dependent variable that depends upon the independent variables x, y, z, and t. Mathematically, then, temperature is averaged over time and space (i.e., global temperature climate) as $\overline{T}(x, y, z, t)$. As of the writing of this book in 2019 our global average temperature is 59 °F.

Where,

$\overline{T}$= Average temperature measured in degrees Fahrenheit (°F) over the entire global atmospheric system space (x, y, z) and a long period of time (t).

x = East to West coordinates of a point on the Earth's land or water surface.

y = North to South coordinates of a point on the Earth's land or water surface.

z = Altitude coordinates from the surface of the Earth to as high as the top of the atmosphere, or as low as the furthest depths of the oceans.

t = time, measured in years, over which the prevailing weather conditions are measured.

With regard to the "z" direction, we typically think of this as the sky, or air, immediately above the surface of the earth.

Meteorologists call this the lower troposphere. But, it can also include the area down to the furthest depths of the seas. The reason for including all the water is that: 1) It covers 71% of the Earth's surface, and 2) Its volume contributes importantly to both weather and climate. The air, land, and sea interface and interaction that generate our weather and climate patterns are especially complex and still poorly understood.

When I'm discussing climate variability in this book, I'm referring to global **climate** and most often to its approximate average temperature ($\overline{T}$).

Occasionally I also will discuss regional, or even local, climate topics. In those instances, I will point out that I have switched to the climates of cities or of the United States, Europe, Asia, Atlantic Ocean, Pacific Oceans, Arctic, Antarctic, Northern Hemisphere, Southern Hemisphere, and so on.

Shades of Blue—Planet Earth's Atmospheric Structure

If you could shrink the Earth down to fit in the palm of your hand, its atmosphere would be as thin as the skin of an apple in Figure 4–2.

Figure 4–2. Atmospheric shell and apple skin comparison. Credit: Grayraven Industries

If you are intrigued by how I came to this conclusion, the math is at Appendix A—Apple-Earth Comparison (see page A–1).

Now we'll look more carefully at the very thin blue spherical, atmospheric shell associated with planet Earth in Figure 4–3. The atmospheric shell is really there, if you look at this picture very closely. Some describe it metaphorically as a "ring" of air surrounding the planet. True. But it is more than that. It is a 60-mile-thick, layered ring of decreasing air density and pressure with height, and it includes layers of oscillating temperature regions.

Figure 4–3. The Earth-atmosphere system.

Notice the tiny black box drawn at the lower right edge of Earth. Although not drawn to scale, the box actually represents a 60-mile up by 25-mile wide cross-section of our atmospheric shell, which is blown up in Figure 4–4, to illustrate the layers within this marvelous shell.

As you can see in this atmospheric cross-section (Figure 4–4),[2, 3] the shell consists of four layers, separated by horizontal red lines. The names of the layers are the troposphere, stratosphere, mesosphere, and thermosphere. The lowest layer, which begins at the surface of the Earth (height = 0 on the vertical y-axis) and goes up to the first horizontal red line, is the Troposphere.

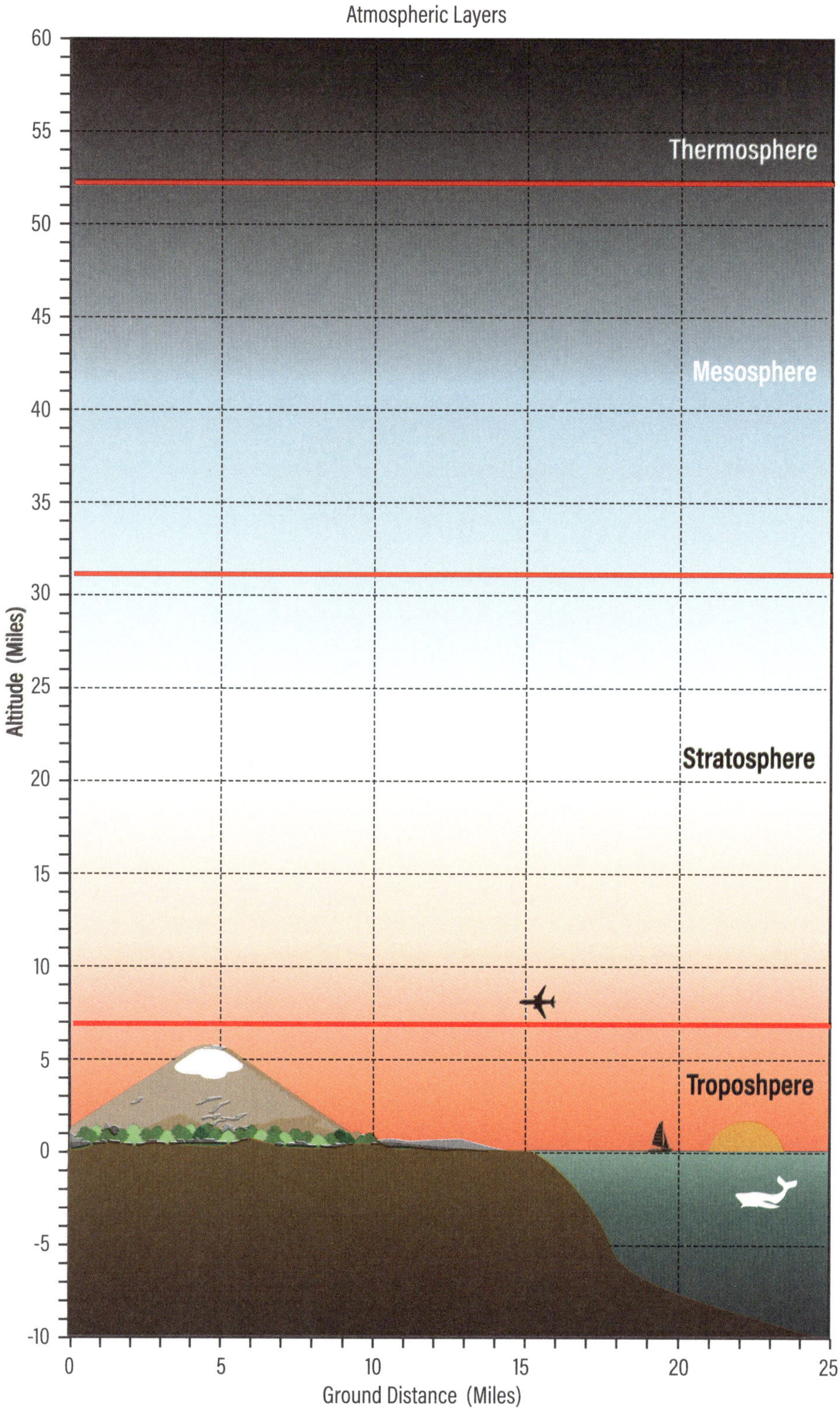

Figure 4-4. Altitude vs. ground distance cross-section of the Earth atmosphere. Credit: Grayraven Graphics

Continuing upward from the top of the Troposphere to the second horizontal red line is the thicker Stratosphere. Then from the top of the Stratosphere to the third red line is the Mesosphere. Finally, the very top portion of this graphic is the Thermosphere. Outer space is everything beyond the 60-mile horizontal boundary at the top, also known as the Karman Line.

At this point you are probably asking yourself a question or two. They might be: Why are the four sub-layers not all the same thickness? What defines the thickness of each of the four sub-layers shown in the diagram? A two-word answer says it all: temperature change. And here's why.

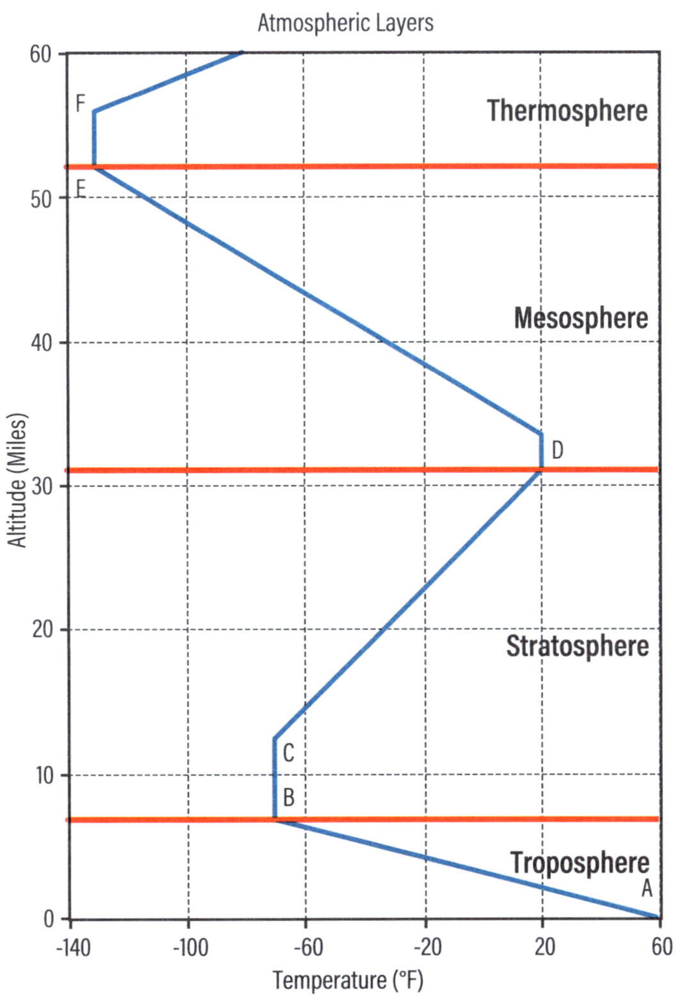

Figure 4-5. Altitude vs. temperature cross-section of the Earth atmosphere.

Figure 4–5 is a bit like Figure 4–4 in that it has a similar altitude scale for the y-axis. But Figure 4–4 is different from Figure 4–4 because the x-axis is now temperature[4] in degrees Fahrenheit (°F), instead of ground distance in (miles). And I have drawn and labeled the segments of the temperature profile as a function of altitude in blue. At ground level, the global temperature is about 59 °F, labeled point "A" on the graphic. As you go upward in altitude, the temperature decreases all the way to minus 70 °F at the top of the Troposphere (point "B" on the plot). The decreasing of temperature with increasing height is called a "negative temperature lapse rate."

Then something strange happens. The temperature remains constant for a little while as you proceed higher in altitude (i.e., constant −70 °F up to point "C"). Then the temperature starts to get warmer as you continue the rest of the way up to the top of the Stratosphere (point "D"). When temperature gets warmer with height, it is called a "positive lapse rate."

Once into the Mesosphere, temperature again remains constant for a small increase in altitude and then rapidly cools with additional increase of altitude, all the way up to the beginning of the Thermosphere (point "E"). Once into the Thermosphere, temperature stabilizes, up to point, and then abruptly changes direction yet again (point "F"), and rapidly warms, with increasing altitude, for the remainder of the Thermosphere. That back and forth "flip-flopping" of temperature gradient is the very property that defines each of the four sub-layers of Earth's atmosphere. It is also the reason for their differing thicknesses.

So, while temperature defines the atmospheric sub-layers, the names of each of the four layers are strictly descriptive in nature.

Tropo means change. This is the layer where 99% of our ever-changing weather occurs; hence the name Troposphere.

Strato means layer. The Stratosphere has a distributed population of air molecules within it, like the other spherical shells. But the

Stratosphere also has a unique internal cloud, or region, called the "ozone layer." This is the good ozone that protects us from incoming cosmic rays.

Meso means middle because temperatures in this layer are approximately in the middle of the layers above and below it.

Thermo means heat, or hot, because it is the final sub-layer where temperature increases with height, and the few remaining air molecules in this layer can get up as high as 4,500 °F.

CHAPTER TAKEAWAYS

1. The atmosphere is really thin. But if you look closely, it has several sub-layers of thickness that are determined by temperature change.
2. All layers of the atmosphere are interconnected and influence each other (e.g., pressure, density, ultraviolet light intensity, etc.).
3. The troposphere is where nearly all the moisture, hence weather and climate, happens.

chapter 5

The Really Big Picture—In Space

Here Comes the Sun—The Thermodynamics of Our Solar System

Have you ever wondered what it's like in the vast expanses of space beyond our Earth and its atmosphere? The yearning to explore the unknown that is above, around, and even below us, is deep-rooted. So, let's now scrutinize our solar system. "System" is a key word here. That's because it is crucial to understand that the sun and the planets and other large bodies of our solar system all **systematically** influence the Earth and its atmosphere and each other. We'll discover how the thermal sources and the inertial and gravity-related forces of the solar system collaborate to be powerful drivers of climate.

Sources

Our solar system encompasses the sun and eight planets from closest to furthest from the sun, namely Mercury, Venus, Earth, Mars, Jupiter, Saturn, Uranus, and Neptune, respectively. Nine planets, if

ACRONYMS USED IN THIS CHAPTER

Acronym	Definition
TOA	Top of (the) Atmosphere
TSI	Total Solar Irradiance in units of W/m^2

UNITS OF MEASURE USED IN THIS CHAPTER

Base Quantity	Unit of Measure	Symbol
Mass	kilogram	kg
Length	meter	m
Time	second	s
Energy	Joule	J
Power	Watts or Joules/second	W or J/s

Base Quantity	Unit of Measure	Symbol
Area	$meter^2$	m^2
Temperature	degrees Celsius or Fahrenheit	°C or °F
Irradiance	Watts per square meter	W/m^2

you choose to include Pluto.[1] Although this solar system is mostly empty space, there are many complex interactions among the planets and other bodies throughout solar space. Each body contributes to Earth ending up in its stable location and orbit. In this section, we'll focus on the specific sun-Earth relationship and the thermal relationship between the two.

Unquestionably, the dominant thermal energy source for the Earth-atmosphere subsystem is the **sun**. From the perspective of size and composition, it is by far the largest object in the solar system, having a radius of approximately 695,700,000 meters (432,288 miles)[2] in comparison to the Earth's average radius of about 6,378,000 meters (3,963 miles).[3] That means the radius of the sun is slightly more than 100 times the radius of the Earth. Also note that the sun is a star and the largest object in the solar system. The sun, along with the four planets furthest from the sun—Jupiter, Saturn, Uranus, and Neptune—are gaseous objects, also known as the "gas giants." Mercury, Venus, Earth, and Mars are planets composed of rock and metal and reside closest to the sun. Of these four rocky planets, only Venus, Earth, and Mars have substantial atmospheric shells surrounding them.

Age-wise, the sun was formed about 4.6 billion years ago, followed closely by the Earth.

Distance-wise, the center of the Earth is about 93,000,000 (93 million) miles away from the center of the sun. In metric units, 93 million miles is the same as 149,600,000,000 meters (149.6 billion) meters away.[3] At this point, I'm going to switch over to metric units for the rest of the book. It's just easier to go metric, because that's how these numbers, especially the energy discussion, are written in the literature.

The sun's radiant power output, combined with the unique distance between the sun and Earth, makes our planet the "Goldilocks" planet, neither too hot nor too cold to support the millions of living species, including humans.[4] The sun generates approximately 382,800,000,000,000,000,000,000,000 (that's 382.8 trillion trillion) joules of heat[5] every second and spreads it outward in all directions. In scientific notation, the sun's radiant power is 3.828 x 10^{26} watts of power (joules of energy per second). I'll be using scientific notation a lot because it is less prone to error, and it is easier for you and me than stringing out all those zeroes.

While 3.828 x 10^{26} watts is a lot of power, only a really small fraction of it—roughly 1,361 watts[5] —reaches the top of the atmosphere (TOA). This is known as the Earth's maximum Total Solar Irradiance (TSI) or **solar constant.** That's TSI at TOA if you are an acronym freak like I am. The rest of the sun's radiation spreads elsewhere beyond Earth and its shadow. Said another way, the sun transmits 3.828 x 10^{26} watts of energy from its entire surface, and the top of the Earth's atmosphere receives 1,361 watts/m^2. A lesser, average irradiance of 340 watts/m^2 bathes the remainder of the surface area of the top of the atmosphere that faces the sun.[6] And, beyond that, the sun's radiation misses the Earth and goes off in other directions. We can think of **radiant power as a transmission** number and **irradiance as a reception** number. [For more details on the solar power road trip, see Appendix B]

Also note that the word "constant" in the term "solar constant" is a little bit misleading because it's not quite constant. It actually varies slightly due to the fact that the sun also has a sizeable number of sunspots, flares, and other activity, that will collectively be referred to as sunspot activity. It provides a small amount of fluctuation to the sun's baseline irradiance.

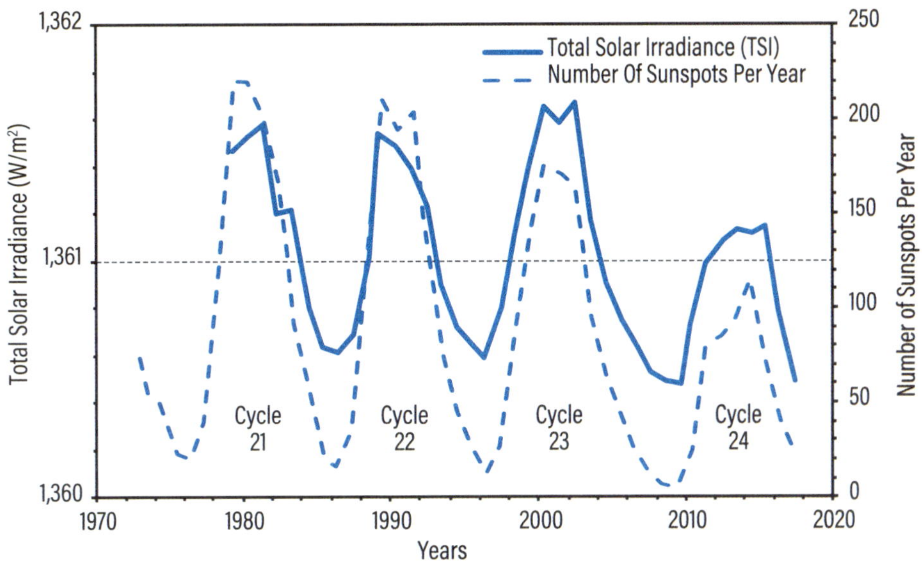

Figure 5-1. Total solar irradiance and sunspot activity.

For comparison, Figure 5-1 illustrates the close correlation among the peaks of TSI at TOA (solid blue curve) and sunspot activity (dashed blue curve) every 11 years or so.[7, 8, 9]

The maximum TSI of the sun is approximately 1,361 W/m^2 at TOA, and the yearly number of sunspots (i.e., sunspot activity) averages between 10 and 220 sunspots. Sunspot activity is cyclic, with maximum values occurring approximately 11 years apart. Numbering of these cycles began at #1 back in 1755, and we are currently nearing the end of sunspot cycle #24. This cycle, which began in 2008 and peaked in 2014, added only about 0.7 W/m^2 or so to the TSI at TOA for cycle #24. However, right now, we're nearing another solar sunspot minimum (the end of cycle #24) expected in late 2019. That additional 0.7 W/m^2 of heat gained over the past 11 years will disappear by the end of 2019, and then we'll be back at the baseline 1,361 W/m^2 of sunlight entering the Earth's atmosphere.

The cyclic rise and fall of the sun's irradiance is quite normal for the sun. And while a sunspot variation of up to 2 W/m^2 may not seem like much, it's more than the two prominent internal energy sources within the Earth, namely nuclear and geothermal sources. My quick back-of-the-envelope calculation suggests that an additional 2 W/m^2 increases the average global temperature by about 0.1 °F, which does not seem significant to me. However, there is an ongoing debate, right now, among climate scientists that sunspot activity, or the change in sunspot activity over time, is a significant player in climate change. That's to be determined.

Attenuation of Solar Energy Coming into Our Atmospheric System

As the maximum of 1,361 W/m^2 of sunlight at the pinnacle of the Earth's atmosphere spreads over the hemisphere of the Earth facing the sun, its average of about 340 W/m^2 descends into the atmospheric shell. The 340 W/m^2 reduces (or attenuates) to about 174 W/m^2 by the time it reaches, and is absorbed by, the Earth's surface. Most of the attenuation is from a combination of scattering, reflection, and absorption of the sun's radiance by the various gases, liquids, and solids in the air between the top of the atmosphere and the Earth's surface. The remainder of the attenuation is due to geometry—the fan-shaped angular spread of solar radiation enroute to the Earth's surface. The sunlight's attenuation-free ride through the vacuum of outer space is over once it hits the TOA.

When all is said and done, the sun's radiation/sunlight/heat translates to an average "global" air temperature at the Earth's surface of about 59 degrees Fahrenheit (°F). That is to say that the sun heats the air, land, and sea near the ground, sustaining the overall comfortable range of temperatures that we experience. Our planet's current global temperature of 59 °F is actually a bit on the chilly side of the historical range of 54–82 °F in the last 542 million years (See Figure 5–11).

The Earth absorbs most of the solar energy arriving at its surface. And the oceans, which cover 71% of the Earth's surface, play a leading role in the distribution of that energy. That's because the oceans are water, a fluid that can move stored heat energy around the globe. So, the oceans are like perpetually moving batteries that store, transport, and release energy over time. This is our second thermal source, called the "oceanic source." And let's not forget that the oceans also deliver water vapor to the air (humidity), some of which ends up becoming clouds and precipitation that ultimately returns to the oceans.

Forces—Inertia, Gravity, and Planetary Motion—The Mechanics

So far, I've described only the thermal process of the sun delivering energy to the atmosphere and the surface of the Earth. Now, we'll switch to the five **mechanical** relationships between the Earth and sun that, for lack of any better words, "manipulate" or "influence" the amount of heat we receive from the sun to sustain life on Earth.

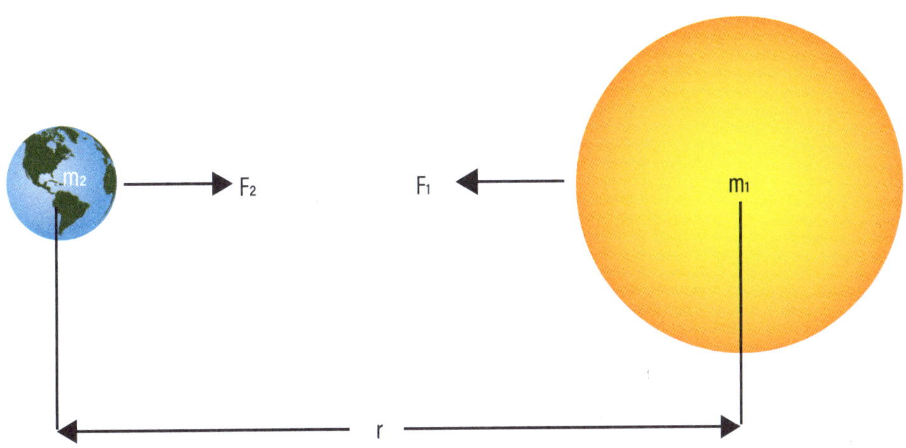

Figure 5-2. Newton's Law of gravitation.

These "force manipulators" are the end result of *Newton's "Universal Law of Gravitation"*[10], as illustrated in Figure 5–2. For our particular application, Newton's law states that the larger sun attracts the smaller Earth toward it. That, in conjunction with the other planets, causes the Earth to orbit, or circle, the sun at an almost fixed orbital distance from the sun. If you are interested in the mathematical description of gravity, see Appendix C.

The important point to grasp in this section is that the gravitational attractive force between the sun and Earth (i.e., the sizes of both, and their distances from each other) keep us in a not too hot and not too cold climate.

Now it's time to add to our list of climate drivers, the five mechanical forces within the solar system that drive our climate:

Force #1—Earth's Rotation About Its Axis

Figure 5-3. Earth's axial rotation.

Figure 5–3 depicts the high frequency, 24-hour counterclockwise rotation, or spin, of the Earth about its axis.[11] As a result, we observers on the ground experience day and night, and the accompanying temperature swings of 10–50 °F that can go with day and night. That's heating and cooling of the Earth, either facing or not facing the sun for several hours. Still, if we average nighttime lows and afternoon highs together across the entire face of the Earth, we have a fairly constant and comfortable temperature of 59 °F.

The importance of the Earth's rotation is the forces that arise from the spinning about its axis. Inertia is the tendency of an object at rest to remain at rest. Or in our case, inertia is the tendency of an object in motion to remain in motion. That's what our Earth has done since its creation, without fail, to this day. It rotates. As the solid portion of the Earth spins, it drags along all the liquid water with it. And the combined motion of land and sea also drag along the air with it. All of this spinning motion of planet Earth contributes profoundly to the ever-changing, but cyclic, behavior of our global climate.

Force #2—Tilt (aka Obliquity)

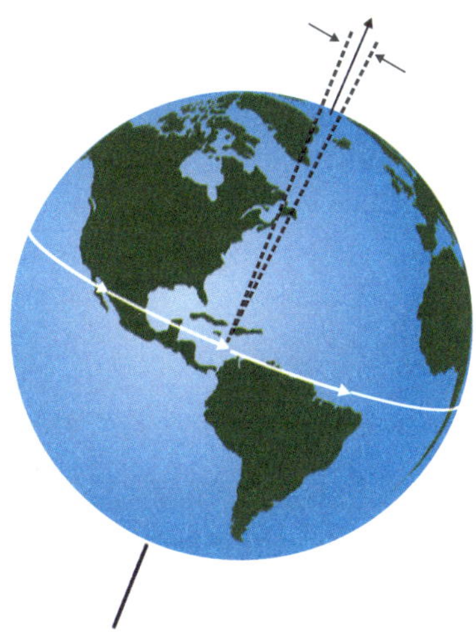

Figure 5-4. The variable tilt of the rotating Earth.

This is the variable **tilt** of the Earth's rotating axis. This tilt is never at 0 °, relative to an imaginary vertical line through the center of the Earth. Rather it varies over a narrow range from a minimum tilt angle of 22.5 ° to a maximum tilt angle of 24.5 ° and then back to 22.5 ° over a 41,000 thousand-year period. Currently, obliquity is right near the median of its travel range, at 23.5 °[12], and headed toward it's minimum value of 22.5 °. And remember that this 2 ° "wiggling" of the tilt is happening very slowly—over a 41,000-year cycle, in comparison to the Earth's 24-hour rotational cycle.

Force #3—Orbital Rotation

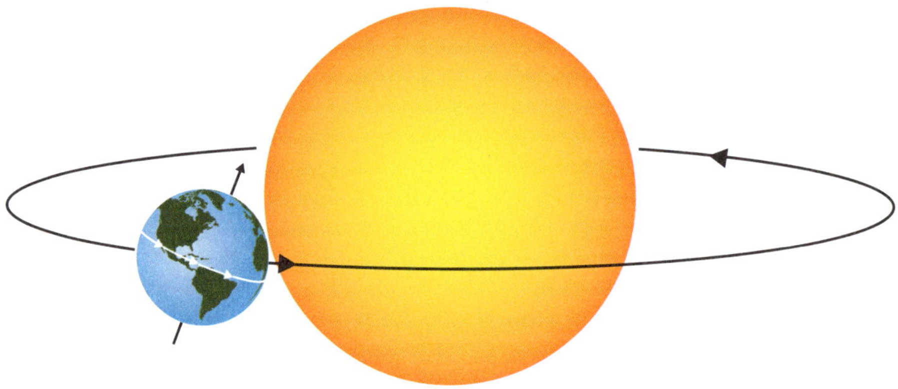

Figure 5-5. Gravity-based orbital motion of the Earth around the sun.

Figure 5–5 shows the spinning and tilted Earth orbiting around the sun once per year (every 365¼ days).[13]

A Note About the Reasons for the Seasons

The seasons of the year are good examples of the complex interactions among the thermal sources and mechanical forces of our solar system. The sun (our dominant energy source) delivers heat energy to the Earth, and the Earth's tilt, orbit, and rotational inertia, are the primary distributors of seasonal amounts of energy to the Earth and its atmosphere. The "reasons for the seasons" are all about two words: "solid geometry." Here's how that works.

Figure 5–6 depicts two separate orbital positions of the Earth as it circles the sun. On both the left and right sides of figure:

The black and blue Earth icons, consisting of segments "A", "B", "C", and "D", represent the Earth at two opposite seasonal positions of its orbit around the sun.

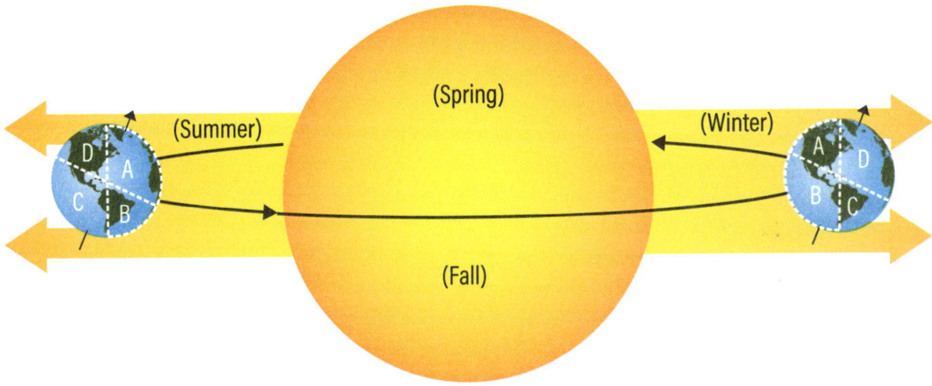

Figure 5-6. Northern hemisphere seasons of the year. Not to scale.

In each case, the tilted black arrows piercing the Earth illustrate the Earth's tilted spin axis.

The diagonal white dashed lines illustrate the equator.

Segments "A" and "D" define the northern hemisphere of the Earth.

Segments "B" and "C" represent the southern hemisphere of the Earth.

The vertical dashed white lines separate daytime segments "A" and "B" from nighttime segments "C" and "D."

Note that on the left side of the figure, the northern hemisphere is in its daytime **summer** season because segment "A" receives significantly **more** sunlight—a larger piece of the heat pie—than its daytime southern hemisphere counterpart, segment B.

When the Earth orbits 180 degrees over six months of time to the right side of Figure 5-6, we have the opposite situation. The right side of the figure illustrates **winter** in the northern hemisphere because segment "A" now receives significantly **less** daytime sunlight than its southern hemisphere counterpart, segment "B." The intermediate orbital positions between summer and winter are where spring and fall occur.

Depending on where you are on the planet, and where the planet is in its orbit, the temperature differential from winter minimum to summer maximum can easily range anywhere from 0 °F to 100 °F.

That's what tilted rotation of Earth around the sun does to climate on an annual basis.

Force #4—Earth's Orbital Eccentricity

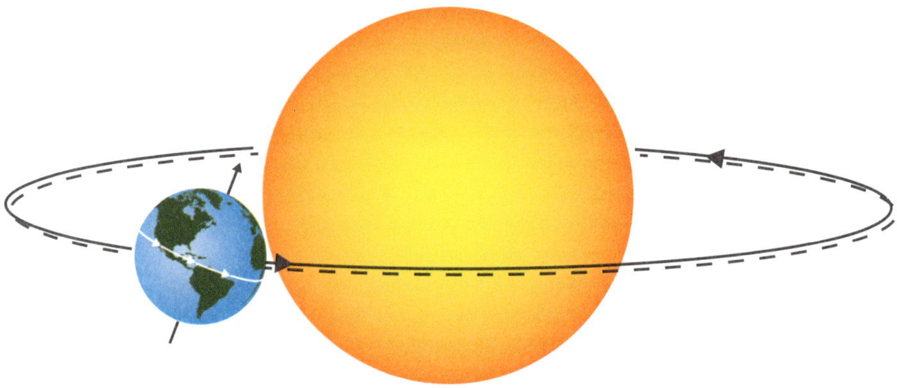

Figure 5-7. Eccentricity of the Earth's orbital motion around the Sun.

The Earth's orbit around the sun, as presented in Figure 5-7, is almost, but not quite, circular. The orbital path is elliptical.[14] We refer to the amount of deviation from a circular pattern as the orbit's "eccentricity." Figure 5-7 depicts that eccentricity by the two

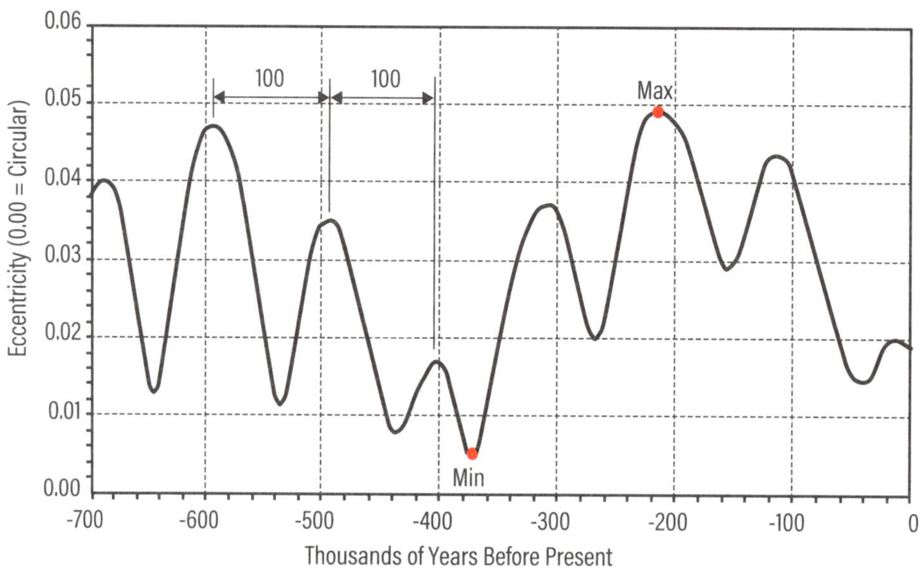

Figure 5-8. Earth's orbital eccentricity during the last 700,000 years.

very slightly differing orbital paths—the solid black maximum and dashed minimum eccentricity paths—of the Earth around the sun.

Figure 5–8 illustrates the small, but varying, amount of eccentricity of Earth's orbit over thousands of years.[15] The vertical y-axis is a measure of how much out-of-round the Earth's orbit is. The horizontal x-axis represents the last 700,000 years, with "0" at the right edge of the x-axis representing today (2019).

The points labeled "Min" and "Max" in Figure 5–8 show when the minimum and maximum values of eccentricity occurred going back 700,000 years.

Several more things are important about Figure 5–8:

- First, Earth's eccentricity from circular varies throughout the 700,000-year period represented in the graph.
- Second, Earth's orbit never gets all the way to a circle, where eccentricity is zero. It gets closest to circular at the point marked "Min" on the graph about 370,000 years ago. At that point, the amount of eccentricity was about 0.005. That's one half of one percent of eccentricity, and very close to circular, zero percent eccentricity.
- Third, Earth's maximum orbital eccentricity is 0.05 (five percent) at the point marked "Max" on the graph about 215,000 years ago. Percentage-wise that's not much eccentricity either. But, as we'll see later on, any of these numbers are more than enough to cause ice ages. Currently (October 2019), the magnitude of Earth's eccentricity is 0.0167.[16]
- Lastly, speaking of ice ages, notice that I have marked the time between three of the peaks of eccentricity on the graph. Each peak is roughly 100,000 years apart. So are all of the surrounding peaks. This is not a coincidence. In fact, the lengths of our historical ice ages also are approximately 100,000 years apart.[17]

The elliptical shape of Earth's orbit and the resultant variation in the distance between sun and Earth do have a warming and cooling effect on the Earth. So do seasonal warming and cooling, which are caused by the tilt of the Earth as it orbits the sun. But, while the variability in the amount of eccentricity is relatively small (5% as noted in Figure 5–8), the difference in the Earth's distance from the sun—three million miles—is quite a lot. See more details about orbital eccentricity of the Earth at Appendix D.

Force #5—Precession

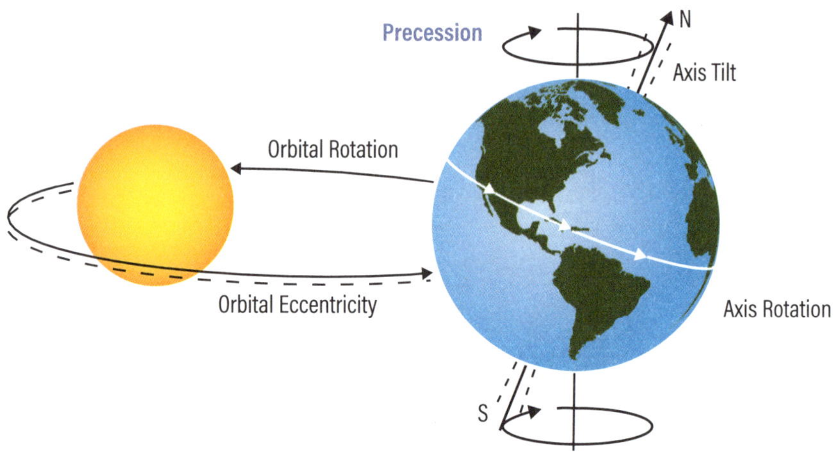

Figure 5-9. Precession, tilt, rotation, orbit, and eccentric orbital motion of the Earth.

Precession is a change in orientation of the Earth's North-South polar axis over time. Precession is the last, but certainly not the least, of the five major, mechanical forces that drive our global climate. The reasons that precession happens are twofold: 1) gravitational pull of the sun and our own moon, and 2) the fact that our Earth is not quite spherical in shape. Actually, the Earth is a bit fatter when measured around the equator than it is when measured around the poles. If the Earth were truly spherical in shape, there would not be any precession.

Figure 5–9 is a not-to-scale illustration of the **clockwise** precession cycle of the Earth's tilted axis around an imaginary vertical line poking through the center of the Earth. Note that this is in contrast to the daily **counterclockwise** axial rotation cycle of the Earth about its polar axis, and the **counterclockwise** orbital rotation of the tilted Earth around the sun.

Speed wise, precession is slow and lazy at a rotational speed of 26,000 years for a complete revolution, or cycle. This is in contrast to a really slow and lazy 100,000-year eccentricity cycle, a still moderately slow and lazy 41,000 year cycle in the tilt of the Earth, and the much faster 365¼ orbital and 24-hour rotational cycles of the Earth. Also, note that precession is thought to coincide with the onset of historical interglacial periods, including the one in which we now reside.[17]

When superimposing all five of these forces that are happening at the same time, many people, including me, find it hard to visualize all this 4-dimensional (x, y, z and time) planetary motion on a 2-dimensional piece of paper. Consequently, I recommend a 12-minute YouTube video to help in this regard. Watch it at this link: https://youtu.be/ZD8THEz18gc.[18]

Powerful, Non-linear, Interaction Among Sources and Forces

The sun and water sources, coupled with the spin, tilt, orbit, eccentricity, and precession forces, regulate the weather of our Earth-atmosphere system at any given moment, and our climate over the long haul. However, they do not act independently of each other. Rather, these climate drivers all interact together in a complex and non-linear manner over small to huge expanses of time.

Figure 5–10 is a recap of the seven powerful, complex, planetary sources and forces that act as "settings" for the temperature profile of our Earth-atmosphere system.

Source/Force (Control Knob)	Periodicity (Cycle Rate)
1. Solar	Continuous (∞)
2. Ocean	Continuous (∞)
3. Spin	24 hours
4. Tilt	41,000 years
5. Orbit	365¼ day
6. Eccentricity	100,000 years
7. Precession	26,000 years

Figure 5-10. List of planetary sources and forces.

Think of these as seven automated control "knobs" on a panel for the heating, ventilation, and air conditioning system in your home that has lasted 4.6 billion years so far.

The first two knobs are fine-tuning knobs. The remaining five knobs are for coarser tuning, and they automatically cycle from minimum to maximum and back to minimum settings at different rates according to their specific function. For example, the "spin" speed knob (symbolizing the rotation of the Earth about its own axis) would cycle one full revolution every 24 hours. The "orbit" speed knob (symbolizing the annual orbit of the Earth around the sun) would cycle one full orbit every 365¼ days. The other three knobs ("Precession", "Tilt", and "Eccentricity") would perform their cyclic motion in the same manner, according to their respective 26,000, 41,000, and 100,000-year rates.

All seven of these knobs have been operating in synchronization since the creation of the Earth. They have produced a temperature profile in our earthly home that looks like Figure 5-11.[19]

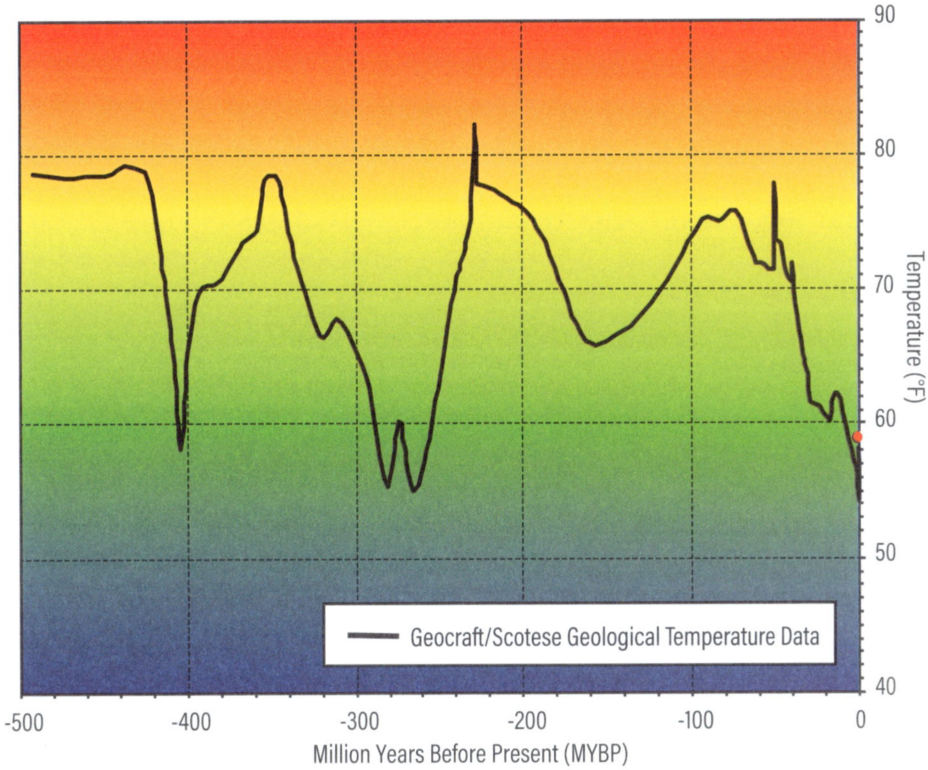

Figure 5-11. Average global temperature during the current Phanerozoic geologic eon.

Notice that the far right-hand edge of the plot ends at today's comfortable, average global temperature of about 59 °F. Also note the large, but controlled, swings in temperature over a very long period of history, for comparison to present day global temperature. This temperature history is thanks to the seven, synchronized and monster sources and forces. The sun, water, rotational inertia, and gravity DO control our climate.

Finally, temperatures have been both hot and cold over the range of the 542 million years shown in Figure 5–11. But, **never** in that expanse of time have temperatures "runaway" in an out-of-control or catastrophic manner. What crystal ball makes us believe that will change in the next twelve years?

CHAPTER TAKEAWAYS

1. There are multiple things happening all at once in this complex, moving, and interactive planetary system in which we reside.
2. The thermal and mechanical sources and forces of our solar system are causes, and our atmosphere and its climate are the effects of these elegant sources and forces.
3. Humans do not control these sources and forces. If humans are causing any degree of "climate change," we produce a mere drop-in-the-bucket in comparison to the powerful thermal and mechanical climate drivers.
4. There is much more to come in later chapters, regarding the influence of all the thermal and mechanical planetary properties on our climate.

chapter 6

The Really Big Picture—In Time

Paleoclimatology—The Historical Climate Record of Planet Earth

History provides a broad perspective about our ever-changing global climate. We'll clearly see very surprising trends that are not at all obvious from our present day-to-day perspective. So, evidence from the past can help with our understanding of the future. Climate models that do not mimic historical data have little chance of predicting the future correctly. But, evidence from the past can help us understand all of Earth's climate timeline.

Geological features are one of the "rear-view mirrors" to past climate conditions that predate satellite-born radiometers and ground-based thermometers. Such groundbreaking techniques (pun intended) for determining temperature and CO_2 levels include ice cores, sediments, fossils, isotopes, and more.

Before taking you on a journey into the past, let's begin with some basic information about the sources and quality of the historical evidence. First, there are four sources of temperature data. They are:

ACRONYMS USED IN THIS CHAPTER

Acronym	Definition
BP or YBP	Before Present or Years Before Present
EPICA	European Program for Ice Coring in Antarctica
HadCRU	Hadley Climatic Research Unit (of the British Meteorological Office)
HadCRUT4	Hadley Climatic Research Unit Temperature (Revision 4)

Acronym	Definition
IPCC	Intergovernmental Panel on Climate Change
MYBP	Million Years Before Present
RSS	Remote Sensing Systems (Corporation)
UAH	University of Alabama, Huntsville

UNITS OF MEASURE USED IN THIS CHAPTER

Base Quantity	Unit of Measure	Symbol
Temperature	degrees Celsius or Fahrenheit	°C or °F

Figure 6-1. Wind River Canyon, Wyoming—Cambrian geological formation. Source: Roy Knorr photo

Thermometers. Since the mid–1800s, the temperature of air, land, and water have been measured with thermometers of one sort or the other. They've been widely used for weather purposes since the late 1800s, and we still use them. Thermometers are point-by-point measuring sticks essentially congregated in the populated areas of the United States, Canada, Europe, Asia, and Australia, but not the unpopulated areas at sea and Arctic and Antarctic regions.[1] From a global perspective they provide coverage of only a small fraction of the surface of the Earth. This is especially true of the oceans and other uninhabited areas.

Satellites. By 1979, satellite-based microwave sounding instruments became operational and supplemented thermometry, but have not replaced it. Collectively, the operational satellites cover virtually the entire surface of the Earth, continually measuring average temperature about 20 times per second.[2] Each measurement covers a roughly a 25,000 cubic kilometer volume of atmosphere in the lower troposphere, near the surface of the Earth.[3] How do these temperatures compare with ground and ocean-based thermometers? Through careful calibration, the ground and ocean-based thermometers and the satellite-based microwave sounding instruments are consistent with each other. I will not be surprised to see satellite instruments take over from thermometers one day. How soon? I have no idea.

Anyway, thermometers and satellites are how we record temperatures today. But we also have several modern scientific methods for discovering what Earth's surface temperatures looked like well before the time of satellite instruments and thermometers. They are primarily ice cores, rock formations, plants, and ocean floor sediments. All can be sampled and analyzed to give us a pretty decent picture of what temperatures have been like in the past, in fact, way back in the past.

Ice cores. Scientists drill deep into ice in Greenland and the Antarctic. They retrieve long, skinny, cylindrical columns of ice, slicing samples from a multitude of closely measured depths, which correspond to different ages. Then they measure carbon and oxygen isotope ratios of air bubbles within the specific samples to determine the temperature of that particular sample. The result is a series of temperature and age pairs for each different-depth ice sample. Ice core sampling has given us temperature history that goes back approximately 800,000 years.[4]

Geophysics/Geology. I lump tectonic motion and all the rock, sediment, plant, and other geology-based sampling into this last category of techniques used to determine temperature history. I won't go into how geophysicists do it, but lets just say that they have been quite successful for many, many decades. As a result of their expertise, we have geophysical temperature proxy records that date back over three billion years.[5]

But, how accurate are these four sources of historic temperature estimates?

Uncertainty, Error Bars, and Margin of Error

Uncertainty is a range of measurements within which the correct or true value is included. Two other terms you hear frequently these days are "error bars" and "margin-of-error." They mean the same thing.

For this book, I made a conscious decision to omit uncertainty estimates from all graphs for two reasons:

Conciseness and clarity of the graphs are more important than the addition of error bars or uncertainty envelopes.

The uncertainties are small enough, in all cases, that they do not change the analysis or conclusions drawn from the data.

For example, in this chapter, you will see temperature data that ranges from today's modern electronic instrumentation, all the way back to geophysical data from 540 million to 3.6 billion years ago. The uncertainty associated with geophysical estimates of temperatures some 540 million years ago is ±2 °F.[6] And, the uncertainty associated with the best of today's temperature-measuring instruments is ±0.5 °F.[7] It's the same stable story with or without the uncertainty envelope.

Global temperature historical evidence has trended downward in a stable manner over the entire temporal range of 540 million years. Moreover, there have **never** been any catastrophic tipping points or uncontrolled runaway temperature events. As a matter of fact, current (2019) temperatures are considerably closer to historic low temperatures, than historic high temperatures.

The story will be similar when we talk about CO_2 (Figures 8–3 and 9–6), tornadoes (Figures 10–5 through 10–14), hurricanes (Figures 11–4, 11–5, and 11–8), etc. There is no big-picture difference in the end result, with or without uncertainty envelopes. So, graphs throughout this book will **not** include uncertainty estimates.

Eyes Wide Open—The Temperature History of Planet Earth

There are those who think that they can predict future temperature patterns and suggest that a climate catastrophe is imminent. Such predictions are unreliable and inaccurate because the climate physics is complex, poorly understood, and not accompanied by real observations and evidence. Cloud physics is a good example of this.

Predictive tools are simply not there yet. Hence, we do need to look in the rear-view mirror for help and understanding. You can't ignore evidence from the past. With that in mind, let's now dive into Earth's climate history to see what we can learn from it, with eyes wide open.

Satellite and Thermometer Data (The Industrial Age)

We begin with most recent evidence about our climate—a profile of Earth's global temperature for the last 40 years in Figure 6–2.

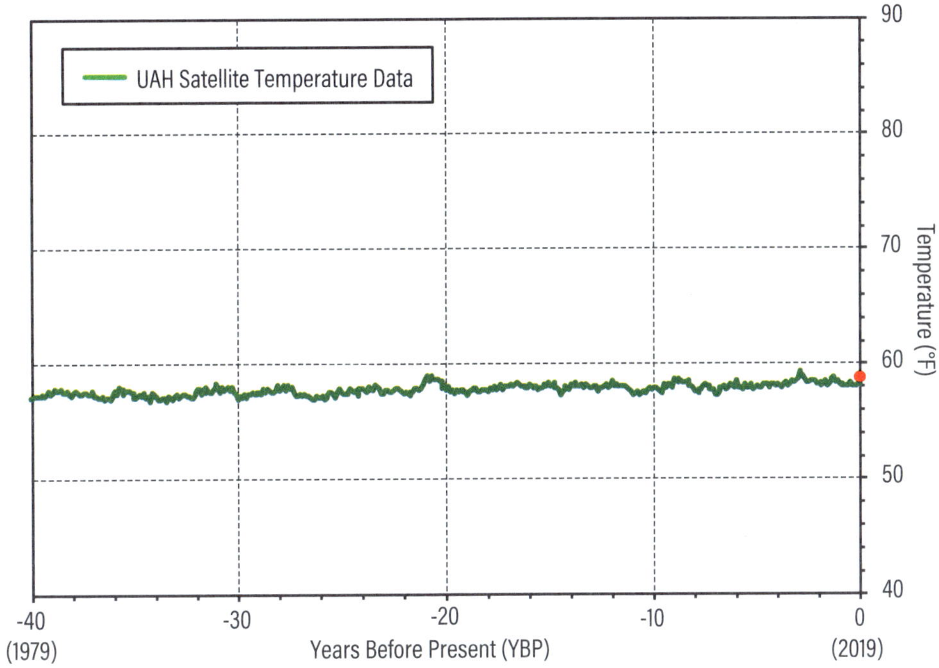

Figure 6-2. Satellite-based global temperature history over the last 40 years

The format for this, and future historical graphs, is a present day (October 2019) zero time value at the right hand edge of the x-axis, with historical time values increasing negatively from right-to-left on the x-axis. (NOTE: If there were any future times in this graph, they would have run positively from zero to the right.) Global average temperature values, in degrees Fahrenheit (°F), are listed as the y-axis, at the right hand edge of the plot. That's because we'll be working from 2019 (year zero) backward in time to 1979 (year –40). The red dot on the right-hand-side y-axis is the present-day global average temperature of 59 °F, as of October 2019.

Figure 6–2 represents a monthly mapping of lower tropospheric (near surface) temperature measurements by microwave sounding units on satellites that map the entire surface of the Earth. Two organizations independently download the data from weather satellites and convert the electronic output of the sensors to temperature measurements near the Earth's surface. The data depicted by the green zigzag line in the above graph are from the University of Alabama at

Huntsville (UAH).[7] The other commercial source, not shown in the plot, is the Remote Sensing Systems (RSS) Corporation. Their temperatures are consistent with UAH's.

These measurements show average global surface temperatures trending upward slightly, at a rate of approximately +2 °F per century. It also shows the highest average global temperature for the 40-year period occurred in 2016 at 59.4 °F. The zigzag nature of the temperature readings demonstrates the combined variability of the radiometers themselves and natural cloud cover, as well as the high resolution of both the 40-year time scale (x-axis) and the 50 °F temperature scale (y-axis). Based on my experience, I estimate the uncertainty of the above measurements to be no more than ± 1 °F. Therefore, the reality of the plot is that actual global temperatures could be anywhere in the range of about 58–60 °F anytime during the last 40 years (1979–2019). This is probably a good time to remind everyone that we're talking about only 2 °F of warming. The smallness of this warming will become more evident in the upcoming graphs, as we zoom out to longer and longer periods of time.

For the remaining series of seven graphs, I will gradually expand the time scale (x-axis) from a **200-year range** before present (BP) to a **four billion year range** BP at the end of the series of graphs. This gives you a feel for how things have changed over the big picture, and shows you that we are currently sitting at a normal, comfortable, temperature of 59 °F. It's a "Goldilocks" temperature that is neither too hot nor too cold when compared to historic global temperature extremes as low as 47 °F to as high as 82 °F.

Regardless of how cold or hot it has been in the past, our planet has managed to flourish quite well. To illustrate this history, the same 40 °F to 90 °F y-axis temperature scale will be used for all of the upcoming graphs. This way, you will be able to easily contrast current temperatures to those of hundreds, thousands, millions, and even billions of years ago.

In the next graph (Figure 6–3), we zoom out from the time span of –40 years (1979–2019) along the x-axis to a time span of –200 years (1819–2019) along the x-axis. In the previous graph, all temperatures were from satellite measurements.

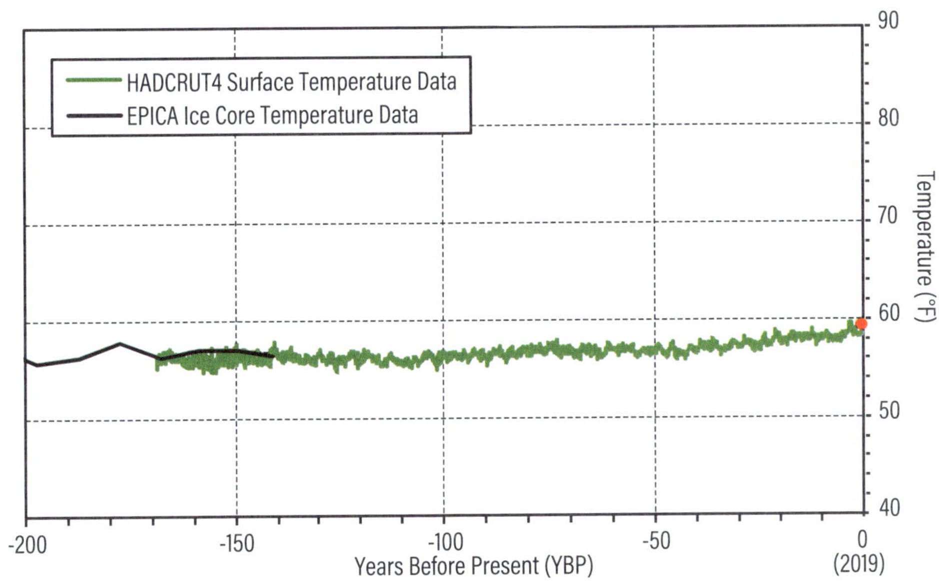

Figure 6–3. Thermometer and ice core global surface temperature historical data over the last 200 years.

In this graph, there are two different sources of temperature measurements that happened during this –200 year time frame. The dark green solid zigzag line represents land and sea surface thermometer measurements by the United Kingdom Meteorological Office's Hadley Center and the University of East Anglia's Climate Research Unit—Temperature (HADCRUT). This is the fourth and latest version of their temperature dataset.[8] The Solid black line data in this graph are temperatures extracted from ice core samples obtained at the "Dome-C" research location in Antarctica.[9]

Moving leftward, backward in time, from the right-hand edge along the solid zigzag line are the HADCRUT measurements. By the way, these land and ocean HADCRUT thermometer measurements closely match satellite measurements from present time back to 40 years ago when weather satellites began operation. The only difference is that satellites record temperatures over the entire globe. Thermometers do not. The satellite record is a considerably more complete record, since thermometers are mostly located in populated regions, and not so much in the polar or ocean regions. In order to compensate for lack of data in remote areas, temperatures from surrounding measurements nearby, and in some cases not so nearby, are interpolated and/or averaged in order to fill in the blanks.

Note that thermometer measurements (the solid green zigzag line) end at about the –165 year ago mark. That's about the time when thermometry technology first became mature enough to use for weather purposes and formal recordkeeping began.

The ice core proxy measurements, represented by the solid black line, now take over to provide temperature evidence further backward in time, and complete this graph at –200 YBP. Note the good interface between ice core and thermometer temperature data where the lines overlap.

These two sources of temperature data continue to show average global surface temperatures trending upward slightly, now at a slightly lesser rate of approximately +2 °F per century, over the entire 200-year period.

Ice Core Data

Figure 6–4 now expands the x-axis to a –2,000 year historical time frame, while keeping the y-axis the same as in Figures 6–2 and 6–3. The reason for keeping the same y-axis temperature scale throughout these graphs will become clear as we proceed further and further back in time in subsequent graphs.

The data sources are the same as in Figure 6–3. However, in Figure 6–4, the instrumental temperature record of thermometers (the solid black zigzag line) now covers only the most recent 160–170 years. And the overlay of EPICA ice core proxy temperature data (the solid green zigzag line) dominates the remainder of the 2000-year time frame.

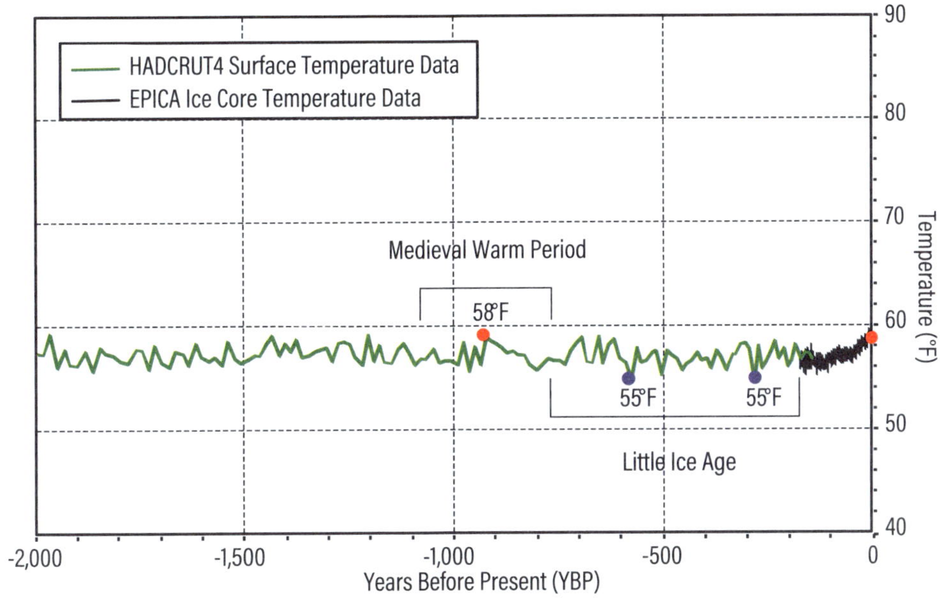

Figure 6–4. Thermometer and ice core historical temperature data over the last 2,000 years.

The temperature profile varies fairly regularly, as expected. But the overall trend throughout is nearly flat. Maximum global temperature increases less than 1 °F from a peak of just over 58 °F during the medieval warm period (MWP) to our current global temperature of slightly under 59 °F. During the MWP, the Vikings inhabited Greenland, and the Black Plague pandemic occurred in Europe.

Also of note during the time frame of Figure 6–4 is what historians refer to as the "Little Ice Age (LIA)." The transition from the MWP to the LIA began around the year 1300 AD and ended around 1850. At its worst, the LIA was about 4 °F colder than today's global temperature of nearly 59 °F. During the LIA, the Vikings abandoned Greenland and London England's river Thames froze solid during the winters.

To summarize: Global temperature is scarcely different throughout this 2,000-year time frame of Figure 6–4.

Now we expand to a –20,000 year time scale in Figure 6–5. This is all EPICA ice-core temperature data, and it illustrates the end of our most recent ice age. (You might want to glance back to Chapter 2 for definitions as you read about ice age terminology.)

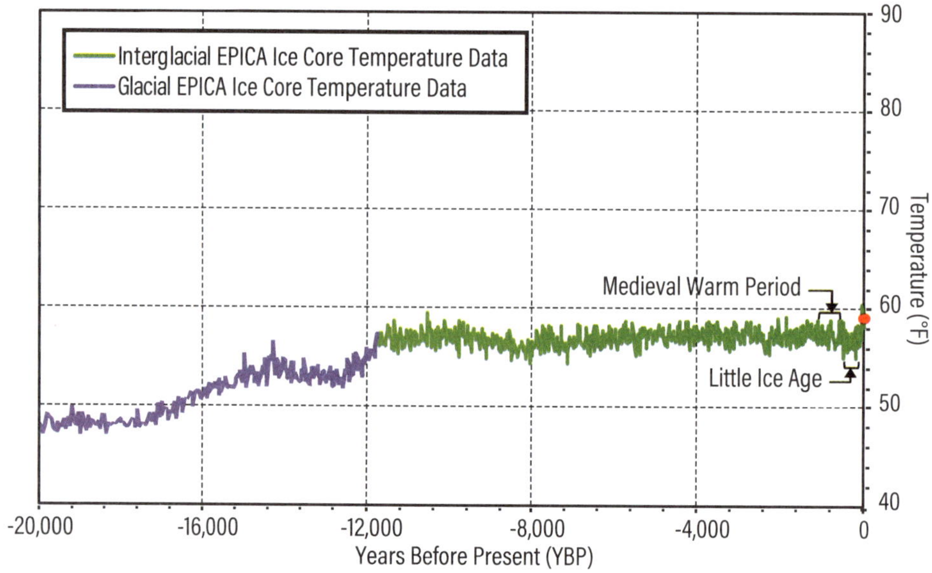

Figure 6–5. Ice core global surface temperature historical data over the past 20,000 years.

The green zigzag line represents the nearly 12,000-year-long interglacial warm period in which we currently reside. An interglacial warm period is a warm period of time **between** cool glacial periods (aka ice ages) on either side of the interglacial. The blue zigzag line represents the ascent from the last ice age, and into our current interglacial period. Once into the interglacial period, global temperature trends stabilized in the temperature range you see.

What caused the end of the last ice age and the rise into our current interglacial warm period? The expansion of our x-axis in Figure 6–6 out to a –200,000 year range of time will help answer the question. It shows our previous and current interglacial warm excursions from otherwise continuous glacial, ice age behavior. Note that the most recent glacial period on the right-hand side of the plot lasts just over 100,000 years.

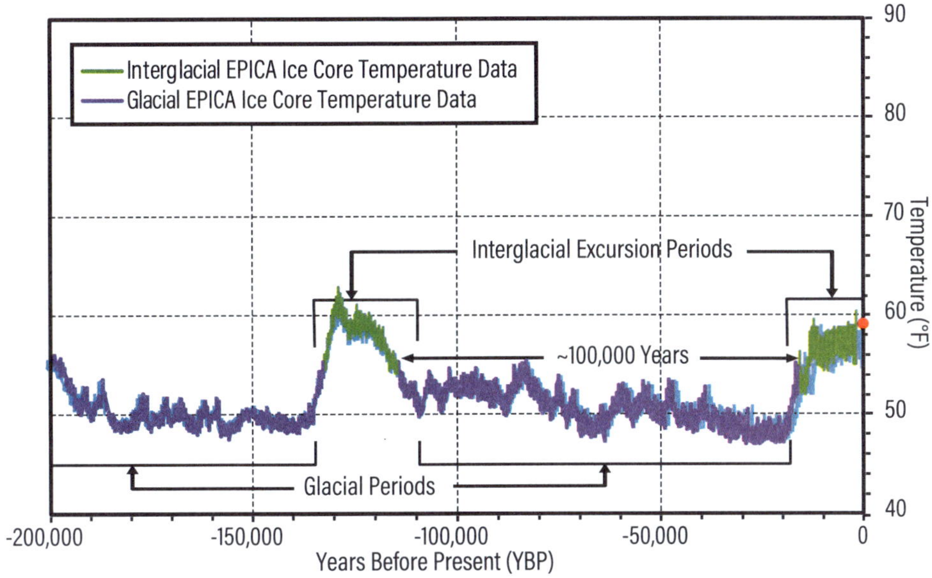

Figure 6-6. Average global temperatures over the last 200,000 years.

Remember, it is the increasing axial tilt and precession of the Earth that we learned about in Chapter 5 that pulls temperatures up and out of the glacial portion of the ice age and into a warm interglacial period of time. Ocean dynamics keep us in warming periods for a while, but then decreasing tilt and precession ultimately release us back down into the omnipresent ice age, where the Earth's orbital eccentricity dominates to keep us there for another 100,000 years or so, before the process repeats itself.

The blue zigzag portions of Figure 6-7 on the next page are successive periods of glaciation. The green peaks are the five interglacial warm periods that occurred during the last 400,000 years before present (YBP) with even more smaller, interglacial reprises going all the way back to –800,000 YBP.

These 800,000 years represent the entire time period for which we have ice core temperature data.

Throughout this 800,000 year period, the average change in global temperature, from highest to lowest, was no more than ~15 °F. That's a global temperature range from 62 °F at the warmest to 47 °F at the coldest. An ice age equals a 10–15 °F drop in temperature. Such a drop is annoying and even painful, to be sure. But it is not deadly to

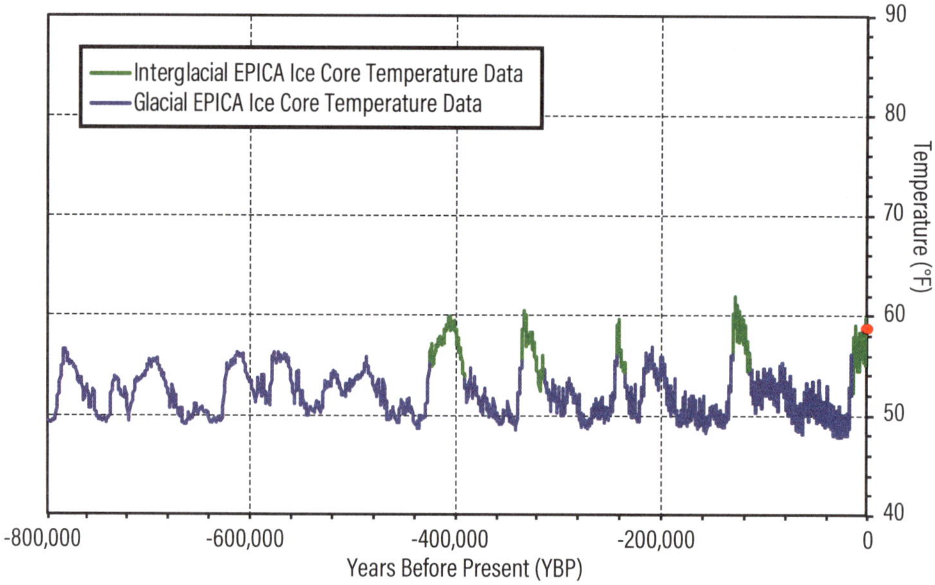

Figure 6-7. Average global temperatures over the last 800,000 years.

the point of human extinction. It is survivable and humans can plan for it.

It means we'll need all the heat sources we can get our hands on, whether be they nuclear, fossil fuel, renewables, hydroelectric and geothermal power, etc. It means we'll need all the food we can store. It means more deaths, because cold kills way more than heat does.[10] And, from a historical perspective, it's nearing the time for another drop into ice age temperatures. It's not a question of whether it will happen, but when it will happen. And no one can say, definitively, when it will happen because we do not understand the physics well enough yet. It may not be pretty, but it will be survivable if we are willing to prepare and adapt. I'll have much more on that in Chapter 9.

Ancient Geophysical History

Now let's look at the upper end of the historical temperature record, which is the very warm topside of the graphs in upcoming Figures 6–8 and 6–9. These two graphs present a further historical journey back in time. These two figures, combined, cover a mind-boggling ~3.6 billion years of highs and lows of global temperature history.[11, 12, 13]

Figure 6–8 is a huge amount of time (542 million years). Yet, it is still only a scant 11% of the Earth's total 4.6 billion-year history.

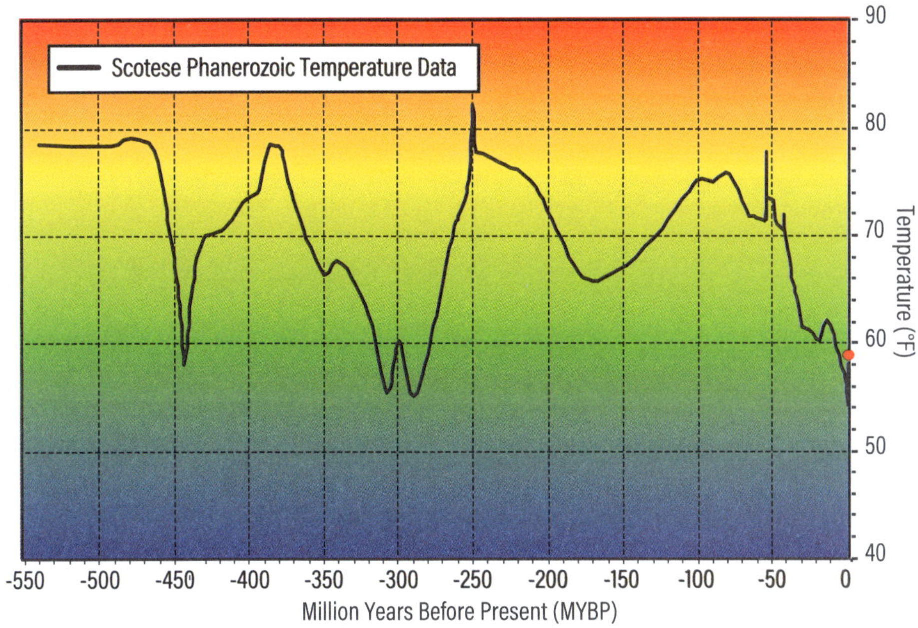

Figure 6–8. Average global temperature during the current Phanerozoic geologic eon.

During this 542 million-year period, we see temperatures again dip into ice ages (blue shaded area) between 290 and 310 million years ago. Also note the long periods of considerably warmer temperatures (red and yellow shaded areas) than our current global temperature. Again note the red dot at 59 °F on the right-hand edge of the plot. The point, here, is that our modern-day temperatures are nearer to historical lows than to historical highs.

Climate **models** generated by the Intergovernmental Panel on Climate Change (IPCC) and others that say we are headed for **catastrophic** warming are not consistent with the ancient evidence shown in Figure 6–8.

As you can see from this graph, we are not now anywhere near either the warmest or the coldest temperature this planet has ever experienced.

In the interest of completeness, now look at Figure 6-9.

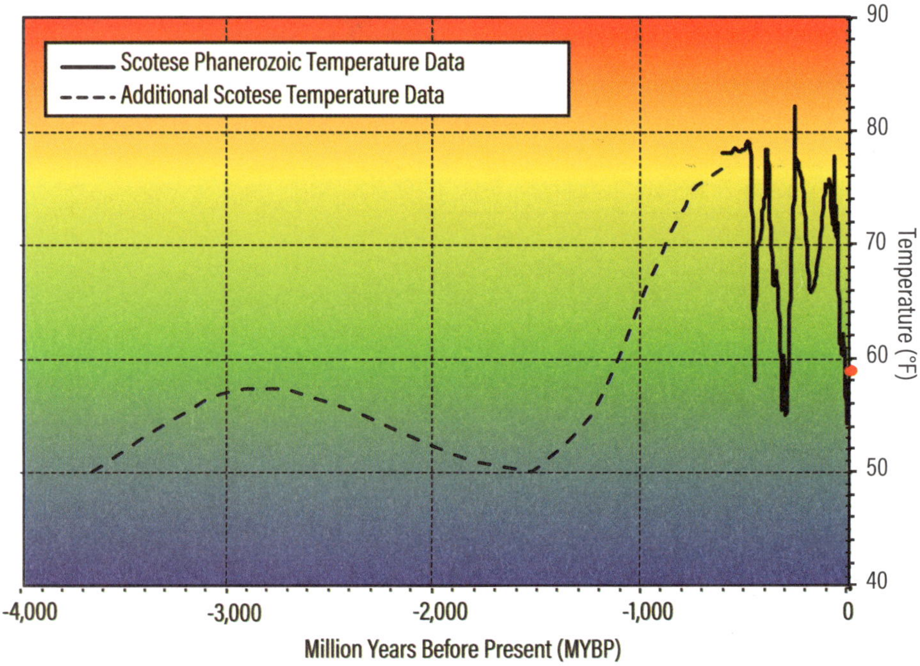

Figure 6-9. Average global temperature during the current Phanerozoic geologic eon..

It is the same plot as the previous one, except this time the x-scale is expanded all the way back to four billion years ago. This graph now shows the earliest known time (roughly 3.6 billion years ago) that geologists and biologists have been able to extract information about temperature history from their various techniques.

The compressed temperature profile version of the most recent 500 million years is the solid line toward the right hand edge of the plot. The dashed line is the rest of Earth's known temperature history back to –3,600 million (i.e., 3.6 billion) years ago. Earth's creation is beyond the left edge of the graph, at 4.6 billion years before present (BYBP). This is the full story on planet Earth's global surface temperature history, as currently known. Earth's earliest 3.6 billion years of existence looks much like the most recent 0.5 billion years in terms of temperature range.

CHAPTER TAKEAWAYS

1. There are four important sources of temperature data that give us insight to the past. They are: thermometry, satellite-based microwave sounding instruments, ice-cores, and geophysical data.
2. Satellite-based temperature profiles are the only source of data that cover the entire planet, thus providing the truest sense of **global** climate variability for the past 40 years.
3. Currently, we live in an ideal, not too hot, not too cold, "Goldilocks" temperature range.
4. Looking back about a million years ago (Figure 6–7) gives an incredible perspective regarding cyclic ice ages, and specifically the later stages of the interglacial warming period that we are in right now. If you believe climate history repeats itself, then you should recognize that extensive cold weather lurks ahead and prepare for it. It's not a question of "if" it will happen, but "when" it will happen. And we are not sure when that will be, because we don't understand all of the underlying physics well enough, yet.
5. Ice age temperatures typically have been 10–12 °F cooler than our current global surface temperature of near 59 °F.
6. Global temperatures also have been much, much warmer—up to 82 °F—for millions of years at a time. But today we are nowhere near those historical temperature extremes. In fact, today we are much closer to historic low temperatures than historic highs.
7. Natural sources and forces (i.e., the sun, water, inertia, and gravity), not humans, cause cyclic hot and cold global temperature swings.
8. Concerned future generations will want to prepare for the next ice age, but how far into the future is unclear.

That's what the really big picture of climate history shows us.

chapter 7

The Air We Breathe

Have you ever wondered what kind of "stuff" is in the air we breathe? Probably yes, especially if you experience smog on a regular basis. To give perspective, let's start with a look at the air here in beautiful, northern Utah, where Nancy and I live (Figure 7–1).[1]

Figure 7-1. Rural northern Utah typical day. Source: Brian Murphy

ACRONYMS AND SYMBOLS USED IN THIS CHAPTER

Acronym	Definition
BAR	Bar(ometric pressure)
EPA	(US) Environmental Protection Agency
GHE	Greenhouse Effect
GHG	Greenhouse Gas
inHg	Inches of Hg (mercury)
μm	Micro (10^{-6}) Meter
μ	Mu: Greek symbol used for volume fraction
PASCAL	International unit of measure of pressure

Acronym	Definition
PM2.5	Particulate Matter (up to 2.5 μm in diameter)
PM10	Particulate Matter (maximum of 10 μm diameter)
PPM	Parts Per Million
PPMV	Parts Per Million by Volume
PPB	Parts Per Billion
φ	Phi: Greek symbol used for relative humidity
PSI	Pounds (per) square inch
UV	Ultraviolet

The Constituents

Let's examine a cubic meter (1 m^3) box of the rural mountain air right near the ground in Midway, Utah. What's in that box of air? A chemical analysis would show that there are thousands of different constituents in the box of air. It's a warm, partly cloudy fall day. So, there isn't any liquid or solid water, i.e., rain, snow, fog, hail, or even clouds in the air right near the ground. The rest of the constituents in the box of air are mostly gases, as well as a small amount of tiny solid particulates of pollution. And all of the gases and particulates are zooming around, continually moving in, out, and throughout the box. That is a snapshot of who is in the box and what they are doing at any point in time on this particular day.

			Name	Symbol	Amount	
Total Constituents in the Box of Air	Clean Air	Gases	Nitrogen gas molecules	N_2	776,734	PPM
			Oxygen gas molecules	O_2	208,369	PPM
			Argon gas atoms	Ar	9,291	PPM
			Carbon dioxide gas molecules	CO_2	370	PPM
			Water vapor molecules	H_2O	5,195	PPM
		Solids	None			
	Polluted Air	Gases	Nitrogen dioxide gas molecules	NO_2	0.005	PPM
			Sulfur dioxide gas molecules	SO_2	0.002	PPM
			Carbon monoxide gas molecules	CO	0.5	PPM
			Ozone gas molecules	O_3	0.025	PPM
		Solids	Soot, smoke, & dust particulates	PM2.5	5	$\mu g/m^3$
			Larger/heavier dust particulates	PM10	10	$\mu g/m^3$

Figure 7-2. Atmospheric constituents in a rural northern Utah box of air.

Figure 7–2 summarizes all the types and amounts, or levels, of constituent players found in our box of air.[2, 3]

The standard way of counting the number of gaseous constituents is parts per million by volume—ppmv, or more commonly just ppm. The ppm designation refers to the ratio, or fraction, of the number of any specific type of gas molecule or atom, relative to the total number of molecules or atoms in the air. For example, I have an instrument that measures the CO_2 level in my backyard. Right now, the display on the instrument reads "370 ppm." It means that out of every one million gas molecules in the air in my backyard, 370 of them are CO_2 molecules, or 370 ppm. If the meter in my backyard were a nitrogen meter, it would say that out of every one million gas molecules in the air 776,734 of them are N_2 molecules, or 776,734 ppm. This applies on down the list of nine rows of gas molecules in Figure 7–2.

However, solid particles in the air are measured differently than gases. Particulates are sorted according to their size range, weighed on itty-bitty balances, and then their concentration is calculated in micrograms per cubic meter ($\mu g/m^3$). The bottom two rows in Figure 7–2 show the particulate pollutants.

Clean Air

In the box of air from Midway, Utah, note that nitrogen and oxygen gases are, by far, the largest number of constituents. Nitrogen, oxygen, and argon together are just over 98% of the total, and are also nearly constant, both globally and over time. N_2 and O_2 do vary slightly to make room for water vapor and carbon dioxide, which are variable gases. Carbon dioxide varies slowly over time. Presently, carbon dioxide is increasing at a rate of about 3 ppm per year, which is a small and beneficial, and not a large and catastrophic, annual increase. In contrast, water vapor can vary as frequently as daily across its entire range of values due to the sun and the effects of temperature and cloud cover.

And lastly, the combined amounts of all the other thousands of gases (e.g., methane, helium, krypton, xenon, and so on) are so small they hardly leave a trace of their existence. Not surprisingly, they are called "trace" gases.

Polluted Air

Now let's look at the pollutants. Like water vapor, the four pollutant gases and two particulates listed in the bottom portion of Figure 7–2 also vary globally, and frequently, over time. But these pollutant constituents depend more upon geography, meteorology, and population than anything else. In the rural, sparsely populated high-mountain valleys of northern Utah, like Midway, the pollutants are very low. As you'll see in upcoming examples, the amount of pollutants tend to be higher in more populated locations like Salt Lake City, Los Angeles, Las Vegas, Beijing, and Bangkok. However, even in heavily polluted locations, the pollutant levels are still small in comparison to the dry gases and water vapor. But make no mistake. Although small in relative abundance, the pollutants make up for the difference with their toxicity. As a result, these small, but potent, bad guys can and do affect health.

Soot and dust are typical particulate pollutants in our little box of air. Unlike gases, particulates are mixtures, or compounds, of solid particles. Soot consists mostly of elemental carbon atoms whose particle sizes are up to 2.5 microns. Wildfire smoke also fits in this category. The formal air quality nomenclature is "Particulate Matter 2.5" or PM2.5. Dust, on the other hand, is a slang term used to describe a conglomeration of larger wind-blown particulates that are a coarser set called Particulate Matter 10, or PM10. The "10" in PM10 refers to the maximum particle sizes in this set, which are between 2.5 and 10 microns in diameter and include oil smoke, and settling dust.

Ozone is different. For starters, it's a gaseous pollutant. It also has a split personality, namely, "good up high, but bad nearby." Ozone that naturally occurs up high in the stratosphere is the good version of this gas. It acts as a sunscreen to protect us from the intense ultraviolet radiation coming from the sun. Stratospheric ozone absorbs UV light up high, by allowing only safer, more subdued, levels of UV to reach the ground.

In contrast, humans also generate ozone near to the ground as a chemical reaction among nitrogen oxides and volatile organic compounds from fossil fuel combustion and heat from the sun. This "nearby" ozone can be a serious health concern at sufficiently high levels, which often happens during the summer months, and even in otherwise pristine air. For example, consider an ozone level of just 86 parts per **billion** (ppb). That's pretty small, yet that amount is sufficiently toxic to be an unhealthy air quality level as defined by the U.S. Environmental Protection Agency (EPA).[4]

In sum, the vast majority of the constituents of air that we breathe are the five most abundant clean gases in the top part of Figure 7–2. They are colorless, odorless, clean, non-toxic, moist air gases. And every one of them (nitrogen, oxygen, argon, **carbon dioxide**, and water vapor) is essential to our life on Earth. The pollutants are the remaining four gases and two particulates in the bottom part of Figure 7–2. They are small in size, but great in toxicity. And each and every one of them (nitrogen dioxide, sulfur dioxide, **carbon monoxide**, ozone, particulate matter 2.5, and particulate matter 10) is harmful to our health.

Besides volume fraction and concentration, there are a variety of other physical properties that characterize the air in our box. These other physical properties include relative humidity, density, temperature, pressure, and so on. For our purposes we'll now analyze the collective results of all the physical properties of our air.

Analysis of Air Properties

Let's analyze the air we breathe by looking at seven different locations on Earth that have fundamental differences in their climate and comparing them to a baseline. The baseline will be clean (no pollutants), dry air, at sea level, and with the current (2019) average global temperature, carbon dioxide (CO_2) volume fraction, and air density. The specifics of those atmospheric conditions are:

- Temperature: 59 °Fahrenheit (15 °Centigrade)
- Pressure: 14.5 psi (29.53 inHg)
- Altitude: 0 feet above mean sea level
- Relative Humidity: 0% (dry air)
- CO_2 Abundance: 410 ppm
- Air Density: 1.225 kg/m^3
- Air Quality: 0 pollutants (clean air)

Also, our top five clean air gases from Figure 7-2 (N_2, O_2, Ar, CO_2, and H_2O) can be treated as "ideal" gases. This means that the atoms or molecules that make up these gases all act like a collection of miniature, billiard balls that are zooming around in our box of air and colliding with and bouncing off each other and not otherwise interacting with each other. The faster they zoom around in our box of air, the higher the temperature of the air. They obey the familiar ideal gas law, $PV = nRT$, which defines the relationships among the physical conditions of pressure (P), volume (V), molecular mass (n), temperature (T), and the universal gas constant (R).[5]

These gases are also odorless, colorless, and non-toxic, and their proportions do not change much over time and space. This is our starting point. These are the conditions of the dominant gaseous players in our atmosphere.

Carbon dioxide and water vapor, besides being two of the top five major constituents, also are greenhouse gases (GHG), whose properties and impacts we'll thoroughly investigate separately in Chapter 8.

Unlike the top four gases, which do not rapidly change in abundance over time, water vapor does. Water vapor can range from 100 ppm in the driest desert climates up to as much as 42,400 ppm in the wettest tropical climates.[6, 7] The amount of H_2O vapor in the air is stated either in terms of the volume fraction of water vapor or relative humidity.

Let's go back to the sample box of air that contains one million parts. These "parts" are the various gases and particulates in the box. For example, on a typical day in rural Midway, Utah, about 776,734 out of any sample of one million (1,000,000) parts in the box are nitrogen gas molecules. Another roughly 5,195 parts of the 1,000,000 parts in the box are water vapor (H_2O) molecules, and so on until we have accounted for all 1,000,000 of the parts in the box, as previously listed in Figure 7–2.

Now, let's focus just on the 5,195 water vapor molecules in the box. We call them the **"abundance" or "volume fraction"** of water vapor parts in the box in units of parts per million (ppm). And, abundance is a good apples-to-apples method for scientists to analyze the various constituents of our air. However, we are more accustomed to hearing or reading weather reports about the amount of water vapor in our air in terms of **relative humidity**, rather than scientists talking and writing about water vapor and other constituents of the air in terms of **volume fraction (aka abundance).**

It would have been nice if relative humidity and the abundance of water vapor were the same numbers. But, the problem is that relative humidity is not equal to abundance. However, while not exactly the same, **abundance** [usually designated by the Greek letter mu (μ)] and **relative humidity** [usually designated by the Greek letter phi (φ)] **are related to each other by a factor "k"**. And, it is important to recognize the difference, namely that factor "k" between them. It has to do with how much water can be in the air before it becomes saturated (i.e., before the air can contain no more water vapor). Mathematically, we represent this concept as:

$$\emptyset = k \cdot \mu = \left(\frac{P_{total}}{P_{Sat}}\right) \cdot \mu$$

Where k is the ratio of the ambient pressure of the air (P_{total}) to the saturation vapor pressure of the water vapor in the air (P_{sat}).

Visually, I'll use the three conceptual boxes in Figure 7–3 to help us even better understand the relationship between relative humidity and abundance. However, keep in mind the conceptual, number crunching nature of the three boxes. By that, I mean that in real life, all the molecules are actually intermingled and in constant motion, not static, separated, and simple to visualize, as they appear in Figure 7–3.

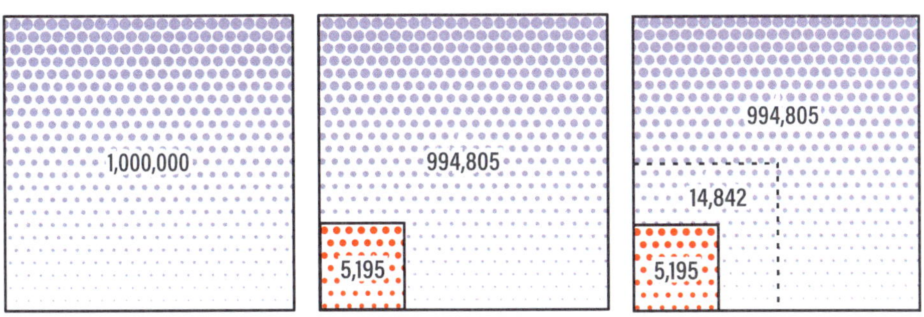

Figure 7-3(a). Total Volume.

Figure 7-3(b). Volume fraction of water vapor.

Figure 7-3(c). Relative humidity of water vapor.

Figure 7-3. Conceptual visualization of the differences among total volume, volume fraction (aka abundance), and relative humidity.

Figure 7–3(a) is our box of air, representing a total of one million gaseous air molecules and atoms of all types (e.g., N_2, O_2, Ar, CO_2, H_2O, SO_2, NO_2, CO, O^3) in the blue box.

Figure 7–3(b) now visually (not physically) separates the water vapor molecules (the red dots from all the rest of the other molecules in the box (the blue dots) according to volume fraction. This is simply the ratio of the actual count of water vapor molecules present (5,195) to the total count of all molecules present (1,000,000). Mathematically, we represent this ratio as:

$$\text{Volume Fraction (Abundance)} = \mu_{H_2O} = \frac{5{,}195}{1{,}000{,}000} = 0.005195 = 5{,}195 \text{ ppm}$$

Figure 7–3(c) also contains the 5,195 parts of water vapor and 994,805 parts of all the other gases in the box totaling 1,000,000 parts. But, this picture also includes a third number—14,842. This is the water vapor saturation level, i.e., the maximum count of water vapor molecules that could be contained in the box outlined by the dashed line. And, it is the ratio of the actual (5,195) to maximum possible (14,842) amount of water vapor in the box that defines relative humidity. Mathematically, we express the relative humidity ratio as:

$$\text{Relative Humidity} = \emptyset = \frac{5{,}195 \text{ ppm}}{14{,}842 \text{ ppm}} = 0.35 = 35\%$$

It is ambient air temperature and pressure that determine the **saturation** number in the denominator of the above ratio. And, the actual amount of water vapor in the box is only 5,195 ppm (35%) of the water vapor needed for saturation. Said another way, any actual water vapor in excess of 14,842 ppm that tries to be in the box will normally condense out as liquid water and fall to the ground. In summary, **just remember that abundance and relative humidity of water vapor in the air are related, but not the same.**

Global Examples of Dry, Moist, and Polluted Air

I've selected a baseline and six different worldwide locations with widely differing climatic conditions for comparison. They are:

1. Baseline: Sea level, dry air and no pollution
2. Midway, Utah: High altitude, low moisture, and low pollution
3. Salt Lake City, Utah: High altitude, low moisture, and sporadic heavy pollution
4. Los Angeles, California: Low altitude, moderate moisture, and heavy pollution
5. Las Vegas, Nevada: Medium altitude, very low moisture, and some pollution
6. Beijing, China: Low altitude, moderate moisture, and heavy pollution
7. Bangkok, Thailand: Low altitude, high (tropical) moisture, and heavy pollution

Figure 7–4 on the next page lists the physical properties of the air for each of the above location scenarios.[6, 7] You can see just how much the numbers differ for high and low humidity, no pollution and heavy pollution, high and low altitude, etc.

For easier reading, Figure 7–4 is labeled by green rows 1 through 22 down the left side and columns A through I across the top. These are the indices of the table. The rows highlighted in blue are major headings of "constituents" and "meteorological conditions." And the rows highlighted in yellow are subheadings of the "constituent" major heading.

- Rows 3–7 list the volume fractions of the five clean gases in the seven atmospheric scenarios of columns C through I.
- Rows 9–12 list the four **gaseous** pollutant air molecules in the seven atmospheric scenarios of columns C through I.

	A	B	C	D	E	F	G	H	I
	CONSTITUENTS			Rural (Midway) Utah	Salt Lake City Utah	Los Angeles California	Las Vegas Nevada	Beijing China	Bangkok Thailand
1	Name	Symbol	Baseline						
2	**Clean Air Gases**		**Volume Fraction (PPM)**						
3	Nitrogen	N_2	780,772	776,734	775,178	769,249	780,488	775,211	752,863
4	Oxygen	O_2	209,452	208,369	207,952	206,361	209,376	207,961	201,966
5	Argon	Ar	9,340	9,291	9,273	9,202	9,336	9,273	9,006
6	Carbon Dioxide	CO_2	410	370	380	404	400	409	410
7	Water Vapor	H_2O	0	5,195	7,096	14,661	257	6,517	35,619
8	**Pollutant Gases**		**Volume Fraction (PPB)**						
9	Nitrogen Dioxide	NO_2	0	5	50	50	85	27	10
10	Sulfur Dioxide	SO_2	0	2	10	20	5	8	18
11	Carbon Monoxide	CO	0	500	700	800	1,100	1,000	1,247
12	Ozone	O_3	0	25	90	90	75	21	22
13	**Pollutant Particles**		**Concentration (μg/m³)**						
14	Soot/smoke	PM2.5	0	5	35	25	35	250	35
15	Heavy Dust	PM10	0	10	60	70	80	354	73
16	METEOROLOGICAL CONDITIONS								
17	Month/Year	mm/yy	10/19	10/19	10/19	10/19	10/19	10/19	10/19
18	Elevation	ft	0	5,700	4,327	144	2,785	180	26
19	Pressure	Pascal	101,325	82,498	86,882	100,814	91,813	100,666	101,234
20	Temperature	°F	59	50	60	75	65	58	85
21	Air Density	kg/m³	1.225	0.972	1.013	1.174	1.076	1.215	1.149
22	Rel Humidity	%	0%	35%	35%	50%	10%	40%	88%

Figure 7-4. Analysis of the constituents of the air we breathe and the meteorological conditions in which they reside in October of 2019.

- Rows 14–15 list the concentrations of the two **particulate** pollutant air molecules in the seven atmospheric scenarios of columns C through I.
- Rows 17–22 are the month/year and pertinent meteorological conditions that apply to each of the seven scenario locations of columns C through I.

Let's look first at a general observation pertaining to the whole table, and then specific observations and analysis regarding each of the seven scenarios.

It is true that the top three atmospheric constituents are relatively constant over time. But, as you read the numbers across the table in rows 3–5, you can see that the volume fractions of even these three highest contributors do vary slightly from location to location. That little bit is normal and expected, due to geography, weather, and pollution level differences among the seven scenarios.

The following scenarios are a sampling of typical clean and polluted atmospheric conditions at different locations on Earth: hot, cold, arid, tropical, high, low, polluted, and pollution-free. This will help to give you an inkling of how air differs from one extreme to another and how local conditions differ from the baseline and one another.

Scenario #1: Baseline (Column C, Figure 7–4)

Baseline refers to nominal conditions in the atmosphere. They are the current global temperature (59 °F) and pressure at mean sea level (1 atmosphere) in clean, dry air, where elevation, pollutants, and relative humidity all are equal to zero, respectively. This is the baseline for comparison with the next six scenarios.

Scenario #2: Midway, Utah (Column D, Figure 7–4)

I picked this scenario because it is a very good example of a nearly pollution-free sample of air nearly all of the time, except on occasion when smoke from wild fires, both near and far, joins our otherwise pristine air. Midway is a scenic, northern Utah, ranching and farming community of approximately 5,000 people, in a high mountain valley setting on the back side of the Wasatch Mountain Range. Its notable meteorological features are its lower humidity, temperature, and air density that result from its higher elevation. What little invisible water vapor and visible pollution that is in this air coexists peacefully and proportionately with the other gases. The Midway box of slightly moist air weighs less than baseline dry air (0.99 kg for Midway vs. 1.225 kg for baseline conditions) for a couple of reasons. One is that the air is less dense at 5,700 ft. elevation than at sea level. As a result, there is less pressure from above pushing or squeezing on the air at 5,700 ft. Therefore, the molecules in Midway are freer to roam than baseline air molecules; hence they are spread further apart (less dense). The second reason is that the lighter water vapor molecules replace a few of the heavier molecules (like nitrogen and oxygen) in the box.

Utah, including Midway, has very dry, arid terrain. Hence, its relative humidity, or volume fraction of water vapor in the air, is usually low. For example, on a sunny, late winter day in Midway, the volume fraction of water vapor is about 5,200 ppm, or 0.52%. That's fairly dry air. Worldwide, the volume fraction of water vapor in the air varies from a low end of about 100 ppm (0.0001%) in desert regions to a high end of 42,154 ppm (4.2%) in tropical regions. Rarely does Utah exceed 1%, i.e., 10,000 ppm of water vapor in the air anywhere in the state. And remember that water vapor volume fraction numbers are not the same as relative humidity levels.

Also note that the CO_2 level of 370 ppm in Midway (Row 6, Column D) is lower than the worldwide sea level average of 410 ppm (Row 6, Column C). This is due to the 5,700-foot elevation of Midway, where the air density is proportionately lower than at sea level. This decrease of CO_2 regionally with height is in contrast to the worldwide and beneficial increase of CO_2 at a rate of 1–3 ppm per year. Fossil fuel combustion, human and animal respiration, etc., produce the CO_2 that is a component of clean air (Row 6, Figure 7–4).

Scenario #3: Salt Lake City, Utah (Column E, Figure 7–4)

Salt Lake City experiences more pollution than Midway, due to Salt Lake's larger size and its valley geography. Truly polluted air happens like clockwork in Salt Lake during certain periods of the year. The typical culprits—NO_2, SO_2, CO, O_3, soot, ash, smoke, and dust—cause the pollution. Their troublesome effects are enhanced, mostly during the winter, as a result of periodic temperature inversions, coupled with the mountainous geography surrounding the Salt Lake valley. The geography traps the pollution horizontally, and the blanket-like temperature inversions (i.e. rising, rather than lowering, temperatures with height) trap the pollution from above. The inversions act as a blanket or cover to hold the pollution close to the ground in the valley. Luckily, this pollution doesn't linger year round, as it does in the remaining scenarios.

When the air is polluted, it is visible, has odor, and can be toxic, in high enough concentrations. The major components of the pollution part of the equation are nitrogen dioxide (NO_2), sulfur dioxide (SO_2), and carbon monoxide (CO) gases, as well as solid particulate matter, namely elemental carbon (C) soot/ash, smoke, and dust. And ozone (O_3) near the surface of the earth is a secondary source of pollution.

All forms of pollution are ugly and harmful. NO_2, SO_2, CO, and C are direct byproducts of fossil fuel combustion by the large number of vehicles, as well as power plants and industry in our society today. SO_2 has trended steadily downward over the years, since the advent of automobile catalytic converters and scrubbers on power plants. NO_2, CO, and C also are headed downward as vehicles are made more efficiently and completely burn their fuel. Due to Midway's rural setting, we have been largely able to dodge the pollution issues of our nearby neighbors in the Salt Lake Valley. But our population is growing.

High O_3 and ammonium nitrate concentrations, resulting from a secondary reaction of sunlight with nitrogen dioxide and volatile organic compounds that come from vehicles, power plants and industry, also prominently contribute to the pollution problem.

Dust often is a windblown, natural pollutant. It also is a human-caused pollutant that results from mining and quarry operations that pulverize our ground and hillsides. A classic example in Utah is the huge, open-pit copper mine on the west side of the Salt Lake valley. This operation is so big it can be seen from space, and it generates plenty of dust.

It is encouraging that most pollution sources in the Salt Lake valley are on a downward trend over the last 50 years due to new technology and clean air rules. Most importantly, the trend in generation of the other two as of yet unmentioned main byproducts of fossil fuel consumption—gaseous H_2O and CO_2—is headed steadily upward for both. H_2O and CO_2 are neither pollutants themselves nor the cause of pollution. That's because, even though they are significant byproducts of fossil fuel combustion, they are colorless, odorless, and non-toxic emissions that are beneficial, and not harmful, to our existence. Life, as we know it cannot survive on Earth without atmospheric water vapor or carbon dioxide.

Scenario #4: Los Angeles, California (Column F, Figure 7–4)

Los Angeles also has a pollution problem that has decreased somewhat over the past several decades. It is also at a much lower altitude than either of the two Utah cities and is more humid. In Los Angeles, geography and temperature inversions are not the main cause. Fossil fuel emissions from energy, industry, and vehicles are the major reason.

But the real problem with pollution in Los Angeles is not so much about pressure and density. It is simply the massive number of people, the enormous number of cars and trucks they drive, and the simultaneously enormous volume of pollutants those vehicles produce. If car and truck manufacturers were to cut pollution emissions by half, that still leaves a tremendous amount of pollution in Los Angeles.

The levels of N_2, O_2, and Ar are quite similar to those of Utah, but humidity and resultant H_2O vapor levels are more than double Utah's levels. That's predominantly a result of Los Angeles being near sea level, and it therefore has a higher pressure and density than the Utah locations. This effectively squeezes water vapor, the other clean gases, and the pollutants more tightly together into a cubic meter sample box compared to the same less-dense box of Utah air.

Scenario #5: Las Vegas, Nevada (Column G, Figure 7–4)

Las Vegas has a significant amount of windblown dust pollution. This is not surprising since it is a Mohave Desert city. The other significant thing about Las Vegas is that it has, by far, the lowest humidity/water vapor content of any of the locations—almost zero. As a result, the levels of N_2, O_2, and Ar are nearly the same as sea level baseline levels, even though Las Vegas is 2,785 feet above sea level.

Scenario #6: Beijing, China (Column H, Figure 7–4)

Beijing has about the same humidity level and altitude as Los Angeles, but is much more heavily polluted by the usual culprits. The soot levels are significantly higher than any of the other scenarios, a result of lingering older and inefficient coal-fired power plants. Even with newer technology coal-fired plants, rumors abound that users circumvent the filters because of the additional cost to run them. Also contributing to Beijing's heavy pollution is wind-blown dust from the Gobi Desert.

Scenario #7: Bangkok, Thailand (Column I, Figure 7–4)

Bangkok is another big city with all the usual big city pollution culprits. Its unique features are its high year-round temperature, as well as its humidity that is nearly to the point of saturation. It is a very moist, tropical city. Bangkok also has a high carbon monoxide component due to older vehicles, factories, and coal-fired power plants, which are less efficient in the burning of fossil fuels.

CHAPTER TAKEAWAYS

1. The magnitude and consistency of nitrogen and oxygen in the atmosphere dominate the air chemistry on our planet.

2. Because oceans cover the majority of the Earth's surface, water vapor content varies widely over space and time. Global water vapor currently averages about 3%, which is on the high side of its worldwide range of 0–4%.

3. Carbon dioxide is not a pollutant. Therefore, it is one of our clean, dry air constituents. It does not contribute to the pollution (smog) that we see frequently in places like Salt Lake City, Los Angeles, and elsewhere.

4. Worldwide CO_2 levels are nearly constant at sea level. The levels decrease at higher elevations. They currently increase slowly and beneficially over time, at an annual rate of approximately 3 ppm per year.

5. Density, temperature, pressure, and relative humidity simply decrease with increasing elevation above sea level in our lower atmosphere (the troposphere).

6. The constituents of the air we breathe vary naturally. They also vary because of human influence.

7. The main pollutants from fossil fuel combustion—nitrogen oxides, sulfur oxides, elemental carbon (soot), and carbon monoxide—have decreased significantly over the last half-century or so.

chapter 8

The "Bad" Guy and "Good" Guy Cases: CO_2 and H_2O

In this chapter, we'll look at two major **internal resources** within Earth's air-land-water system—carbon dioxide (CO_2) and water (H_2O)—and how they interact with the major **external resources**—gravity, inertia, and the sun—to produce the food we eat, the water we drink, and the comfortably warm, life-sustaining, "Goldilocks" climate that we experience 24/7/365.

What Would We Do Without Them?

What would we do without H_2O and CO_2? The answer is easy for H_2O. Without water, life cannot exist. The answer is similar for CO_2. Without CO_2, food cannot be grown. Without food, life cannot exist.

To understand why this is the case, let's start with a revisit of the "big five" atmospheric gases in Figure 8-1. All five are clean gases, meaning none of them are pollutants. And since water vapor is one of the five, we refer to them as a clean humid mixture of gases.

ACRONYMS USED IN THIS CHAPTER

Acronym	Definition
°C or °F	Celsius or Fahrenheit (degrees)
GHE	Greenhouse Effect
GHG	Greenhouse gas
LWIR	Long Wave Infrared (radiation)

Acronym	Definition
NASA	National Aeronautics and Sapce Administration
NOAA	National Oceanic and Atmospheric Administration
PPM	Parts Per Million
VOC	Volatile Organic Compound

SYMBOLS USED IN THIS CHAPTER

Symbol	Definition
Ar	Argon Atom
CH_4	Methane (Gas Molecule)
$C_6H_{12}O_6$	Glucose (Molecule)
CO_2	Carbon Dioxide (Gas Molecule)
H_2O	Water (Gas, Liquid, or Solid Molecule)

Symbol	Definition
N_2	Nitrogen (Gas Molecule)
N_2O	Nitrous Oxide (Gas Molecule)
O_2	Oxygen (Gas Molecule)
O_3	Ozone (Gas Molecule)

Figure 8–1 is a typical coastal example of our atmosphere with about 757,490 molecules of nitrogen gas (N_2), 203,100 molecules of oxygen gas (O_2), 30,000 water vapor molecules (H_2O), 9,000 argon atoms (Ar), and 410 molecules of carbon dioxide (CO_2). The two columns on the right hand side are the quantities, or percentages of parts, in a volume containing a total of one million parts. In this example, the parts are the number of molecules and atoms in column 3, or the percentage of molecules and atoms in column 4.

GLOBAL ATMOSPHERIC COMPOSITION (LOWER TROPOSPHERE)			
		Humid Volume Fraction	
Constituent	Symbol	(PPM)	(%)
Nitrogen	N_2	757,490	75.7%
Oxygen	O_2	203,100	20.3%
Water Vapor	H_2O	30,000	3.0%
Argon	Ar	9,000	0.9%
Carbon Dioxide	CO_2	410	0.04%
	Sum:	**1,000,000**	**100.0%**

Figure 8-1. Global atmospheric composition in the lower troposphere.

CO_2 is the least abundant of the top five atmospheric gases at 410 ppm. Yet it certainly gets the most notoriety of the big five. And it seems almost all of the press is bad.

The common story line to the public is:

1. The combustion of fossil fuels is a major producer of energy for vehicles, homes, industry, and more—**TRUE.**
2. CO_2 is a major byproduct of the fossil fuel combustion process—**TRUE.**
3. Not surprisingly, the amount of CO_2 in the air is rising—also TRUE.
4. CO_2 is a greenhouse gas—also **TRUE.**
5. Greenhouse gases absorb heat, some of which would have otherwise exhausted to outer space—also **TRUE.**

Conclusion: Therefore, CO_2 is a pollutant gas that causes smog in the air, and is a major reason for catastrophic levels of global warming—**FALSE.**

This is the climate-alarmist narrative that floods our society these days. The first five statements, above, are correct. But the sixth statement—the conclusion—is categorically false. The conclusion does not follow from the first five statements for several "devil-in-the-detail" reasons.

A few of the important truths that debunk the conclusion, above, are:

1. CO_2 is a colorless, odorless, non-toxic clean gas. It is not a pollutant that causes health issues. Nor does CO_2 mix with other gases or particulates to become a pollutant.
2. CO_2 is the breath of life for intake by plants that then convert it to produce food we eat and oxygen in the air we breathe.
3. Water vapor, not CO_2, is the dominant greenhouse gas that redirects enough heat back to Earth to keep our global temperature at a variable, but comfortable, level to sustain **all life.**

4. Except for a tiny amount of heat captured by CO_2 in the greenhouse effect, it is actually changes in temperature that drive changes in CO_2 levels, and not the other way around.

Historical and current scientific evidence gathered in earlier climates supports the above statements. Let's now continue the discussion of our cyclic, but well-controlled Earth-atmosphere system by focusing on energy balance and the forces and processes that affect energy balance.

Conservation of Energy in the Earth/ Atmosphere System

The principle of energy conservation has to do with the total energy of an isolated system, such as our Earth/atmosphere system, which must remain constant, or conserved, over time. This means energy can neither be created nor destroyed. But, the Earth/atmosphere system can still warm and cool without violating the principle of conservation of energy. More on that later in this chapter.

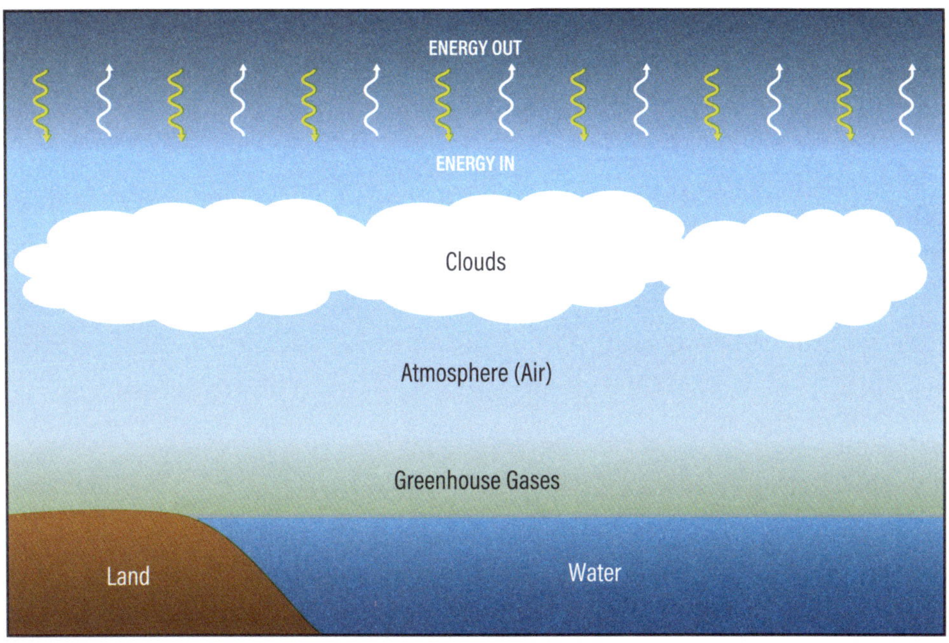

Figure 8-2. Notional diagram of the energy balance in Earth's air-land-water system (not-to-scale).

NOTE: In the interest of brevity and clarity, from here on out I will simply use the phrase "the system", "our system", or the "Earth's air-land-water system" to mean an invisible "box" that contains the Earth's crust and water to the deepest point of any ocean and the entirety of its surrounding atmospheric shell.

Figure 8–2, is a notional, not-to-scale, diagram of "the system." Besides air, land, and water, including clouds, the diagram also depicts all the solar energy that enters the system from outer space, as well as all the solar and terrestrial energy that exits the system back to outer space.

If the temperature at the Earth's surface were constant over time, then the energy coming in from the sun to the system would equal the energy going back out of the system to the vacuum of outer space. But, we know that is not the case. Air temperature is not constant. It varies up and down over time. When our air is warming globally, more energy from the sun is coming in than going out. Similarly, when we are cooling globally, more energy is going out than coming in. It's a bookkeeping drill of "ins and outs" to and from the system. Nevertheless, while the energy doesn't always balance between in and out, not one iota of energy can mysteriously appear or disappear yesterday, today, or in the future. Solar energy is all accounted for somewhere within our Earth-atmosphere system. Note that I was specific in the previous sentence when I wrote, "Solar energy is all accounted for . . ." There are a couple of other energy sources that are internal to the Earth, namely tectonic and volcanic energy. Both are actually the results of gravity and the spin of the Earth about its axis. Energy can be tricky that way. Fundamental sources of energy, such as the sun, inertia, and gravity, can and do morph from one form to another within our system, but there still is no overall gain or loss of energy in the process.

Radiation (the movement or propagation of the sunlight that we see and the heat that we feel) is the dominant form of energy that pertains to our system. Incoming solar radiation has a specific range of wavelengths (mostly visible colors) associated with it. In contrast, outgoing terrestrial radiation has a different wavelength range of invisible (infrared) colors associated with it. The incoming and outgoing forms of radiation are fundamentally different in color, amount of energy, and temperature. However, from a "system perspective," no energy is gained or lost. That's because there are four more forms of energy sharing within the system. They are convection, latent heat, conduction, and feedback, which will be covered in more detail in Chapter 9. Just remember, from the first to the last light, none of the energy of the day is lost during the shuffle. Keeping track of which photon is where is an entirely different story.

Let's begin with a description of the individual pieces of the big picture and why they are important.

The CO_2 Big Picture

Life in general, and our life-sustaining Earth system in particular, is really all about the big, contextual and complete picture, and not cherry-picked, disjointed, incomplete pieces of a picture. That's why it is necessary to be open-minded in the view of the picture in both time (history) and space. And by "space," I mean the unabridged air-land-water system we live in, not just a local or regional piece of it, nor the vast connotation of "outer space." Our Earth system is, to be sure, only a piece of our vastly larger universe. But it is still a "relatively big picture" entity, in its own right, that has performed miraculously throughout its 4.6 billion year history.

Never has a "big picture" mindset been more important than in understanding carbon dioxide. Patrick Moore, founding member and former president of Greenpeace, articulates: "CO_2 is the most important nutrient of all life on Earth."[1] Without it, our planet would be uninhabitable. What is presently worrisome is not that there is too much of it in the atmosphere, but rather too little of it. **Right now, CO_2 is near its lowest level in 538 million years. That is near starvation level.**

With the preceding in mind as starting points, let's take a closer look at carbon dioxide.

Carbon Dioxide (CO_2) History

Figure 8–3 shows CO_2 history over the last 530 million years.

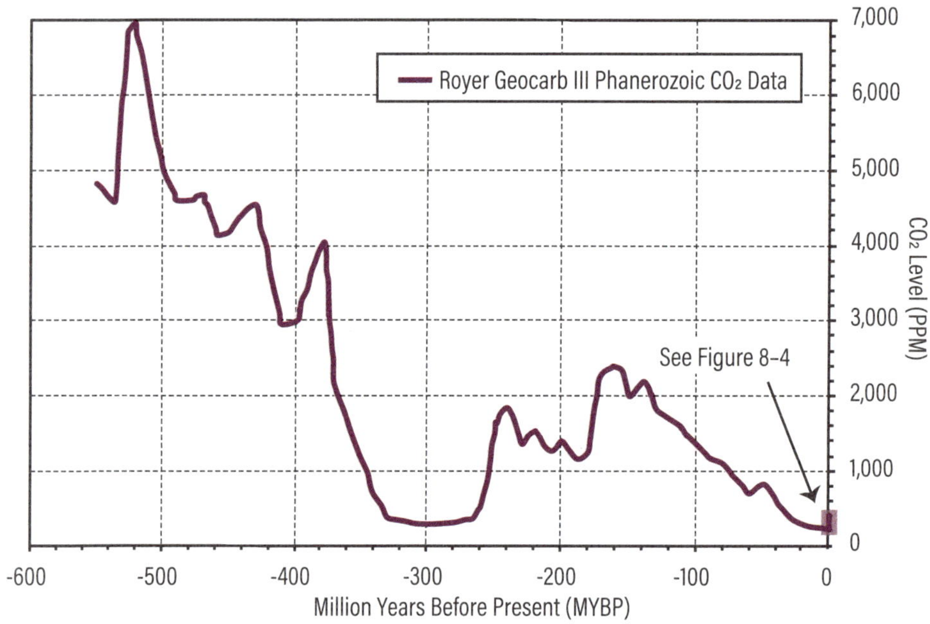

Figure 8–3. CO_2 concentration from 0 to –600 million years before present.

The purple curve is CO_2 levels that peaked at nearly 7,000 ppm, some 530 million years ago (near the top left side of the graph), and stretch all the way down and over to our present level of approximately 410 ppm today (2019).[2]

In contrast, during our most recent 800 thousand years, indicated by the "tiny purple box" in the lower right-hand corner of the graph, our airborne CO_2 levels have ranged from an all-time low of about 188 ppm to today's 410 ppm.[3] But even at 410 ppm, that's a far cry from around 7,000 ppm of CO_2 in the air. Figure 8–3 demonstrates that modern CO_2 levels are at near **all time lows, not highs**, relative to the complete historical record.

Some people claim that the high levels of CO_2 in the early days are irrelevant to today's situation. I say: No. The early values are just as important as the current levels. An intricate path of growth has led to our current, comfortable climate. And considering the exponential growth of today's world population, more people need to be fed. The way to do that is via a more efficient photosynthesis process that results directly from the availability of more, not less, CO_2 and water.

The purple curve in this graph is evidence from a variety of sources, including biological and geological data that range from fossil plants, coral, rock erosion, ocean sedimentation, radioactivity, plate tectonics, and ice core data.

Now let's take a closer look at an expanded view of that tiny purple box from Figure 8–3.

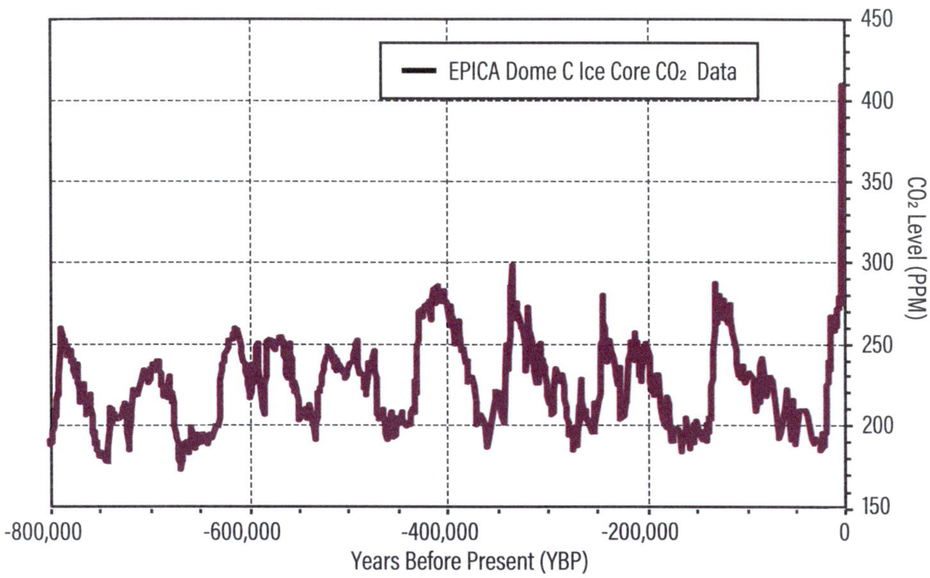

Figure 8–4. NOAA climate.gov graph of CO_2 over the last 800,000 years.

Figure 8–4, is an expanded (800,000 year) view of the tiny purple box portion in the lower right hand corner of Figure 8–3 on page 130. It is also an oft-cited 800,000-year NOAA graph[3,4] that climate alarmists use in their "playbook" to scare people into thinking that CO_2 is now accelerating out of control to record high levels. Well, it's not.

They interpret this NOAA graph to mean that CO_2 levels have been below 300 ppm until very recently. But because of their conveniently expanded x and y-axes of the graph, the CO_2 level appears to jump upward quite rapidly in recent time, highlighting our fossil fuel boom. They also say that this supposedly sudden and exponential rise in the CO_2 level is driving global temperatures higher, right along with the CO_2, due to the greenhouse effect. Case closed. The "science is settled," so they say.

The reality is that our recent CO_2 increase is quite mild when viewed in a complete big picture mode. Their ignominious message "cherry picks" 0.1% and ignores 99.9% of the available data that goes all the way back to 551 million years ago. As such, it hides evidence that, even with the recent uptick in CO_2 concentration, we are nearing **starvation-low** levels, not **record-high** levels. And the use of a mere 0.1% of available data by critics is hardly what I would consider "record-worthy."

Also, saying that CO_2 drives temperature upward gets cause and effect backward. What we do know from the totality of the previous two graphs is that much higher early-on CO_2 levels sometimes occurred during periods of ice age cold temperatures. CO_2 abundance has decreased by more than a factor of ten in the last 550 million years. And if you overlay 550 million years worth of temperature and CO_2 levels on the same graph and look closely, it is clear that changing CO_2 does not drive changing temperature. In fact, geophysical records show that temperature leads CO_2 at all measured time scales. Modern instrumentation shows that temperature leads CO_2 by 9–12 months.[5, 6]

More conspicuous evidence that CO_2 changes **follow** temperature changes, which also happens to be one that is close to my "Cokeaholic" heart, is that carbonated soft drinks do not hold their CO_2 on a hot summer day. They go flat because they release their CO_2 **after being warmed**, not before. And they do it quite rapidly at that. Any drinker of carbonated beverages knows this. Changing CO_2 does not cause changing temperature.

Still, climate critics argue the opposite. They contend that CO_2 is a greenhouse gas (true). And greenhouse gases trap some heat that would otherwise escape to outer space. Therefore, CO_2, a greenhouse gas, drives temperature (false). When all is said and done and we look at the complete picture, the big picture, we see that the "big dogs"—the sun, water, inertia, and gravity—rule the "kennel."

The data is misused by climate alarmists. Dr. John Christy, Alabama's State Climatologist, and Distinguished Professor of Atmospheric Science at the University of Alabama, Huntsville, exposed this when he pointed out misuse of data by alarmists. He said: "They (alarmists) are not aware of what goes into it (the data) . . . Here we have a science that is so dominated by personalities that claim the science is settled, yet when you walk up to them and say prove it, they can't."[7]

It is true that currently CO_2 is rising, and has been doing so at a **steady, non-accelerating** rate over the last 200 years. But in the big picture sense, CO_2 is near its lowest level since Earth's creation 4.6 billion years ago, which is my definition of "ever" for recordkeeping purposes. And that is just outdoor CO_2 levels.

Indoor vs. Outdoor CO_2

Outdoor CO_2 levels are uniformly distributed and currently at 410 ppm, compared to 280 ppm about 100 years ago, and 7,000 ppm about 550 million years ago. We humans, like it or not, do spend a lot of time indoors. What are CO_2 indoor levels like today? For comparison, Figure 8–5 provides a few examples of typical CO_2 levels **inside** some really interesting places.

These common CO_2 concentrations can vary widely, depending on ventilation characteristics and the number of people involved (i.e., the amount of CO_2 they exhale). For example, I performed the CO_2 measurements at my home, a local church, a multistory office building,

Location Type	Examples of Indoor CO_2 Concentrations (PPM)
Homes	900
Churches	1,500
Multistory Office Buildings	1,800
Apollo 7 Space Capsule	3,000
Classrooms	4,000
International Space Station	5,000
U.S. Navy Submarine	8,000
Space Shuttle	10,000
The CO_2 We Exhale	40,000

Figure 8-5. Examples of indoor CO_2 levels.

and a classroom. NASA Astronaut Walt Cunningham (of Apollo 7 and Skylab fame) provided the Apollo 7 Space Capsule, International Space Station, U.S. Navy Submarine, and Space Shuttle examples.[8]

We commonly breathe in some pretty high levels of CO_2 that are neither toxic nor dangerous. CO_2 is invisible, and has no odor whatsoever. It does not contribute to any health or respiratory issues. What is so bad, or unhealthy, about inhaling 410 ppm of CO_2 inside or outside? Nothing.

That's reassuring, but also ask yourself: What good does CO_2 do for me, anyway? The answer to this question is pretty important. First, CO_2 feeds the plants that feed us. It's an airborne, nontoxic fertilizer. Second, CO_2, and its partner H_2O, generate most of the oxygen that is in the air we breathe via the process of photosynthesis. And third, CO_2, along with water vapor, helps redirect some heat that would otherwise escape back to space—the "Greenhouse Effect." If it were not for those greenhouse gases, our Earth's temperature would be quite chilly all the time.

Water—The Big Picture

As mentioned earlier, water is pretty straightforward. It is also quite unique in the Earth-atmosphere system. Water (H_2O) is the only widespread molecule that exists at the same time in all three of its states—solid, liquid, and gas.

As a gas on a worldwide scale, H_2O is the third most abundant contributor to the air we breathe (Figure 8–6).

GLOBAL ATMOSPHERIC COMPOSITION (LOWER TROPOSPHERE)			
Constituent	Symbol	Humid Volume Fraction (PPMV)	(%)
Nitrogen	N_2	757,490	75.7%
Oxygen	O_2	203,100	20.3%
Water Vapor	H_2O	30,000	3.0%
Argon	Ar	9,000	0.9%
Carbon Dioxide	CO_2	410	0.04%
	Sum:	1,000,000	100.0%

Figure 8-6. Global atmospheric composition in the lower troposphere.

However, unlike the relatively fixed quantities of nitrogen, oxygen, and argon, water vapor is a variable gas, ranging from 0 to 4% by volume and globally averaging approximately 3% by volume.[9]

Water vapor also occupies the atmosphere in its liquid form as clouds, rain, and fog, and in its solid form as ice and snow crystals. On the surface of the Earth, it's either liquid water, or solid ice or snow. Our oceans, which cover 71% of the Earth's surface, are also huge reservoirs of both heat and CO_2. Additionally, we know that H_2O is essential to all life on the Earth, be it human, animal, fish, bird, or plant life.

Most importantly, H_2O and CO_2 work together to generate most of the oxygen we breathe and food we eat. How so?

Photosynthesis

The first way, or process, is photosynthesis. Figure 8–7 describes the chemical process.

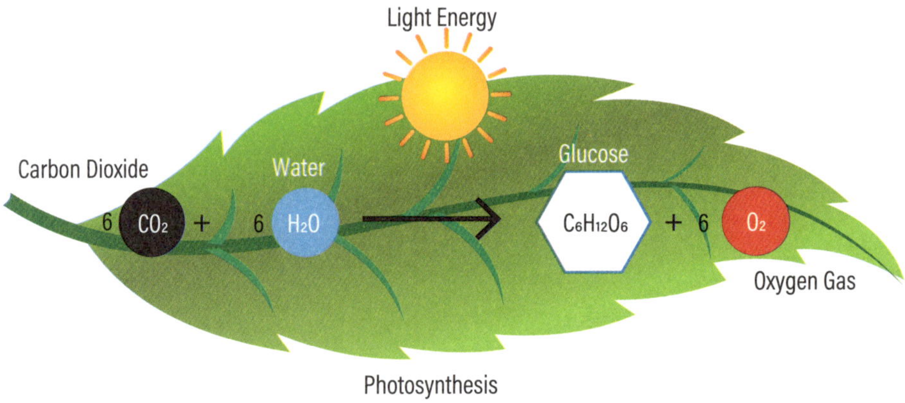

Figure 8-7. The chemistry of photosynthesis.

Figure 8–8 illustrates the process of photosynthesis in plants. Solar radiation energizes vegetation to take water up through its roots, as well as air through pores (aka stomatal apertures) in leaves, stems, etc. A portion of the water (H_2O) and carbon dioxide (CO_2) feed the plants, generating glucose ($C_6H_{12}O_6$), a specific type of carbohydrate as one product of photosynthesis. The other product of the photosynthesis process is oxygen (O_2). The stomatal apertures serve to exhale unused air and water in vapor form, and the

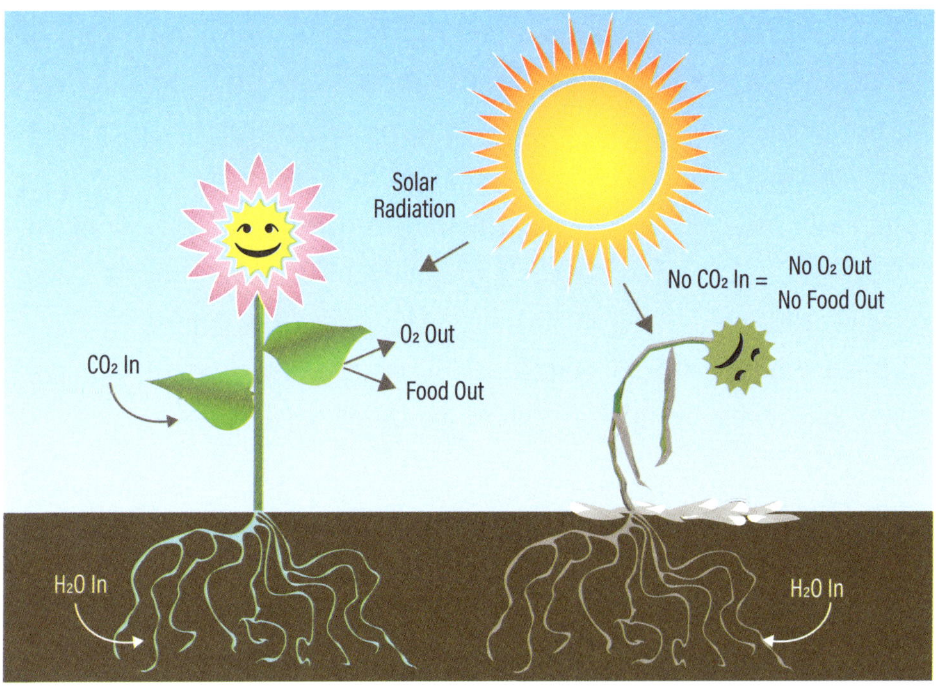

Figure 8-8. The land-based photosynthesis process.

photosynthesis product O_2 out to the atmosphere. In other words, vegetation breathes just as humans and animals do.

The carbohydrates become food we humans eat, and the oxygen becomes about 98% of the oxygen in the air we breathe.[10, 11, 12] As illustrated above, land vegetation, like rain forests, grows well in the presence of sufficient H_2O, CO_2 and sunlight. Marine vegetation, like seaweed and kelp, does even better, accounting for more than 50% of the oxygen in the air.

On the other hand, insufficient CO_2 stunts growth, regardless of the availability of water and sunlight. And so the whole process of photosynthesis collapses, taking the oxygen and food sources down with it.

On a more positive note, there are a couple of bonus effects of plentiful CO_2 that are not captured in the diagrams. One bonus is increased plant growth rates, or yield, which is the reason growers like to pump extra CO_2 into their greenhouses. Another bonus is that enhanced CO_2 allows many types of plants to grow better in arid, or even drought conditions, than they would otherwise.

For the last 50 years, the Dutch have grown tulips in intense CO_2 environments. The result was that the additional CO_2 stimulated better growth of these flowers. And this stimulation is true for nearly all plant life. Carbon dioxide in the atmosphere is a powerful fertilizer. Independent proof positive is an experiment that Dr. Sherwood Idso performed, in the USA, on four pine trees of identical age and size. He grew the pine trees in greenhouses with four differing levels of CO_2, with the results as shown in Figure 8–9.[13]

At the conclusion of this experiment in 1994, the outdoor ambient (AMB) CO_2 level was 385 ppm. The four CO_2 levels tested are, from left to right, 385 ppm, 535 ppm, 685 ppm, and 835 ppm. The increased pine tree growth resulted from increased CO_2 as a fertilizer in the presence of H_2O, yielding increased growth as well as photosynthesis products, namely carbohydrates and oxygen. Carbon dioxide is the ingredient that sustains nearly all vegetation on the surface and in the sea. The more CO_2 that vegetation sucks in,

Figure 8-9. Growth of pine trees in ambient and enhanced carbon dioxide conditions. Source: Sherwood Idso.

the bigger and better it grows. The bigger the vegetation, the more grains, vegetables, fruits, and trees we humans have for food to eat, fibers for clothing, and lumber for shelter.

The Greenhouse Effect (GHE)

Besides photosynthesis, the other physical phenomenon that has a major effect on our air-land-water system, and specifically our climate, is the greenhouse effect (GHE). Simply stated, the Earth "cools at night." Cooling means the Earth radiates heat that it absorbed from the sun during daylight back toward outer space at night as "terrestrial radiation." But the greenhouse gases in the air, absorb some of this escaping heat and redirect, or feed it back, toward the Earth. That "feedback" of heat energy is the GHE, and it keeps the Earth warmer than it would otherwise be at night, without the presence of greenhouse gases (GHGs) in our air.

Figure 8–10 illustrates the major greenhouse and non-greenhouse gases, ranked by volume fraction near the surface of the Earth.

The GHGs are highlighted in yellow. Water vapor is, by far, the most abundant GHG, at an average **global** volume fraction of

AVERAGE GLOBAL COMPOSITION OF THE AIR NEAR SEA LEVEL			
		Volume Fraction	
Gas Constituent	Symbol	(PPM)	(%)
Nitrogen	N_2	757,488	75.7%
Oxygen	O_2	203,100	20.3%
Water Vapor	H_2O	30,000	3.0%
Argon	Ar	9,000	0.9%
Carbon Dioxide	CO_2	410	0.04%
Methane	CH_4	2	0.00%
Tropospheric Ozone	O_3	0.30	0.00%
Nitrous Oxide	N_2O	0.01	0.00%
Other GHG Trace Gases	CFCs & VOCs	0.01	0.00%
Other Non-GHG Trace Gases	Many	0.002	0.00%
Total of all gases:		1,000,000	100.0%

Figure 8-10. Volume fraction of greenhouse gases and non-greenhouse gases.

approximately 30,000 parts per million (ppm), or 3%[14] of the total gases in the air nearest the surface of the Earth. Carbon dioxide, currently at approximately 410 ppm by volume, or 0.04% of air, is a distant second place GHG, followed by the remaining GHG trace gases at a collective 2.5 ppm.

The greenhouse gases are present in the air all the time. But the GHGs only absorb outgoing terrestrial radiation, and not incoming solar radiation. That's because solar and terrestrial radiation emit at distinctly different wavelengths. The GHGs absorb at wavelengths that match up with the outgoing terrestrial radiation emissions, and not the mostly visible incoming solar radiation emissions. The band of GHG absorbing wavelengths is in the longwave infrared (LWIR) region of the electromagnetic spectrum, which, unlike sunlight, is invisible to the human eye.

The GHE begins when the Earth's surface begins to cool each day by emitting LWIR terrestrial radiation. This cooling begins before sundown, typically toward late afternoon, and lasts until sunrise. The primary atmospheric gases from Figure 8–10 (nitrogen, oxygen, and argon) do not interact with the upward-bound LWIR terrestrial radiation. But the greenhouse gases certainly do interact. This begins the blanket-like, heat-trapping effect of the GHGs as they absorb and then re-emit (i.e., "catch and release") a substantial amount of outbound LWIR radiation back down to the surface of the Earth. This is the greenhouse effect, and it does have a dramatic effect on global temperature at the Earth's surface—a dramatically good effect!

Without any GHGs in the air, our global temperature would be approximately −1 °F.[15] That's 33 °F below freezing, and a full 60 °F lower than our actual global temperature of +59 °F. So, the GHGs, led by water vapor, make life comfortable and hospitable on our planet. They merely help contain a whopping 60 °F worth of exiting energy (−1 °F to +59 °F) that would otherwise travel back to outer space unused.

Here is one more general thing about the greenhouse effect. It's not new. It's been around since the creation of the Earth and not just since humans began using fossil fuels. The greenhouse effect has existed since water and carbon dioxide have been in the air.

In the next chapter, I'll continue with more on the effects of H_2O, CO_2, and the greenhouse effect on the Earth's overall energy balance.

CHAPTER TAKEAWAYS

1. Everything about CO_2 and H_2O is good for us. Together, they are fuels that generate the oxygen we breathe and the food we eat. It's that simple. Yes, some fractions of them are also byproducts of fossil fuel combustion. But neither are pollutants that need to be taxed or eliminated. The focus should be on the real culprits: Volatile Organic Compounds (VOCs), ozone (nearby, not up high), and particulates (smoke and dust). None of these has anything, whatsoever, to do with clear, odorless, non-toxic CO_2.

2. Photosynthesis and the greenhouse effect are two vital processes that utilize two vital molecules, H_2O and CO_2, to support all life on Earth—producing the food we eat and the oxygen we breathe.

3. Carbon dioxide has not before caused, and is not now causing, runaway, catastrophic warming of the atmosphere. Multiple ice ages that followed after 7,000 ppm levels of CO_2 are clear evidence of that.

4. We know that a greenhouse gas, like CO_2, should warm the Earth, and it does to a small degree. But, in the bigger, more complete, scheme of things that include the sun, water, inertia, and gravity, it is changing temperature that drives changing CO_2.

5. CO_2 does not drive warming. It prevents freezing.

chapter 9

Energy Balance and the Greenhouse Effect

Energy Accounting 101

Energy comes into our Earth-atmosphere system, and energy leaves the system, mostly equal in and equal out, but not always. When more comes in than goes out, we warm. When more goes out than comes in, we cool. And energy can be transformed from one form to another. But, one thing is certain. Energy in a closed system, like our solar system and our Earth-atmosphere system, is conserved. Energy is **neither** created nor destroyed. That's conservation of energy.

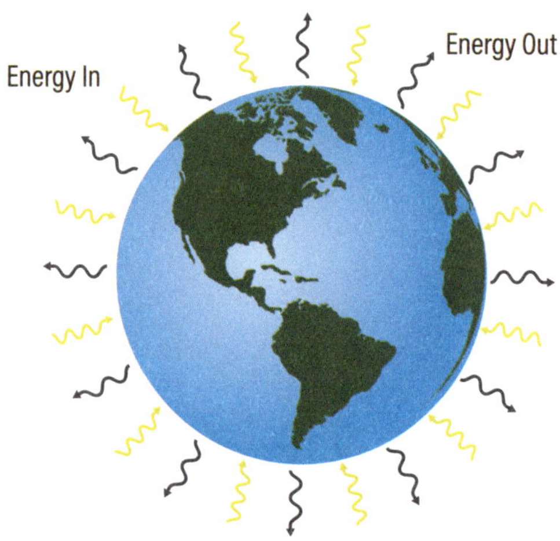

ACRONYMS USED IN THIS CHAPTER

Acronym	Definition
GHE	Greenhouse Effect
GHG	Greenhouse gas
LWIR	Long Wave Infrared (radiation)
MYBP	Million Years Before Present

UNITS OF MEASURE USED IN THIS CHAPTER

Base Quantity	Unit of Measure	Symbol
Area	meter^2	m^2
Energy	Joules	J
Irradiance	Watts per square meter	W/m^2
Power	Watts = Joules/second	W = J/s
Temperature	degrees Celsius or Fahrenheit	°C or °F
Volume Fraction	Parts Per Million or Parts Per Million By Volume	PPM or PPMV

Now we are ready to run the actual numbers on movement of energy from the sun down to and through our air-land-water system and back out to space. This chapter illustrates the conservation of energy and the transformation of energy, with emphasis on the role of the greenhouse effect.

We begin with a total solar irradiance (TSI) discussion. TSI is " . . . the **maximum** possible power that the sun can deliver to the Earth at its average distance from the sun."[1] TSI is written in units of power, or Watts (W), per square meter of surface area of the Earth (W/m^2), and its value at, and perpendicular to, the Earth's surface is approximately 1,361 W/m^2. However, when " . . . **averaged** over the entire planet, the amount of usable sunlight arriving at the top of Earth's atmosphere is only one-fourth of the total solar irradiance,"[1] or approximately 340 W/m^2 that bathes the Earth.

For our Earth's energy/power (think temperature) to be conserved, the amount of energy exiting must equal the amount coming in. Otherwise, our Earth system would be unstable, and would have burned up or frozen solid long ago. The reason it hasn't is periodic warming and cooling cycles that maintain the long-term average of

340.3 W/m^2 both in and out of the system. (If you are interested in where this number comes from, see Appendix B, for the details.)

Figure 9–1 is a summary breakdown of solar energy as it enters, travels within, and exits the air-land-water system[2]:

SOLAR ENERGY AUDIT—SUMMARY		
Description	Irradiance Fraction (%)	Irradiance Value (w/m^2)
Reflection Back To Space—Subtotal:	**30%**	**102.1**
By cloud tops	20%	68.1
By atmospheric aerosols	6%	20.4
By Earth's surface	4%	13.6
Absorption Within The Earth—Subtotal:	**70%**	**238.2**
By atmosphere	19%	64.7
By Eath's surface subtotal	51%	173.6
Re-emit LWIR directly to space	6%	20.4
Latent heat of H_2O condensation	23%	78.3
Convection	7%	23.8
Greenhouse gases	15%	51.0
Total Reflection + Absorption	100%	340.3

Figure 9-1. Solar energy audit summary.

Here's more detail about what happens to the energy in Figure 9–1.

- **The net of 30%** of reflected solar energy travels toward outer space.
- **The net of 70%** of absorbed solar energy, as itemized below, is re-emitted as LWIR radiation that is ultimately sent back to outer space:
  - **19%** of LWIR re-emitted by **atmospheric gases.**
  - **6%** of LWIR re-emitted by **land and water.**
  - **23%** of LWIR re-emitted as **latent heat of condensation of rising water vapor.** It ultimately condenses to liquid cloud droplets and transfers that energy to the surrounding atmosphere. Subsequently, these nearby atmospheric gases re-emit their gains as LWIR toward outer space as the gases cool off.

- **7%** of LWIR re-emitted by land and water goes to **convective (rising air) transfer of energy to surrounding air** and ultimate re-emission of gains as LWIR toward outer space as the air cools off.
- **15%** of LWIR re-emitted by land and water is **absorbed by greenhouse gases.** That energy is then re-emitted as LWIR radiation in all directions. About half goes away from the Earth, and the other half goes toward the surface of the Earth, and is absorbed as bonus heating. This is the Greenhouse Effect (GHE). As the Earth rotates about its axis, it releases this 7.5% of bonus heat as it cools. And, like everything else, that energy also ultimately makes its way toward outer space as LWIR energy.

The entire list illustrates the approximate balanced energy (irradiance) budget of 340 W/m^2 in to the system ≈ 340 W/m^2 out of the system. But it's that last 7.5% of the 15%—The Greenhouse Effect (GHE)—which is the focus of so much consternation these days. This is where the dominant greenhouse gas (GHG)—water vapor—does its job to maintain our current 59 °F global surface temperature.

At this point, we'll now shift gears and focus on that last 15% of the terrestrial energy (radiation) that the greenhouse gases capture, and what happens to that energy and how it happens.

Figure 9–2 illustrates the spectral difference between incoming solar radiation (yellow curve) and outgoing terrestrial radiation (red curve).[3] The important point of this graph is that solar radiation and terrestrial radiation have different emission colors, i.e., wavelengths (x-axis values). That shift in wavelength from visible incoming solar radiation (yellow curve) to outgoing LWIR terrestrial radiation (red curve) is why the greenhouse effect exists.

The greenhouse gases absorb only the outbound terrestrial LWIR and re-emit about half of the absorbed energy back toward Earth. Hence, the GHGs serve as a "blanket" to redirect otherwise escaping LWIR and preserve warmth of the Earth at its comfortable global temperature of +59 °F. If the greenhouse gases were not present, Earth's temperature would be a much chillier −1 °F.

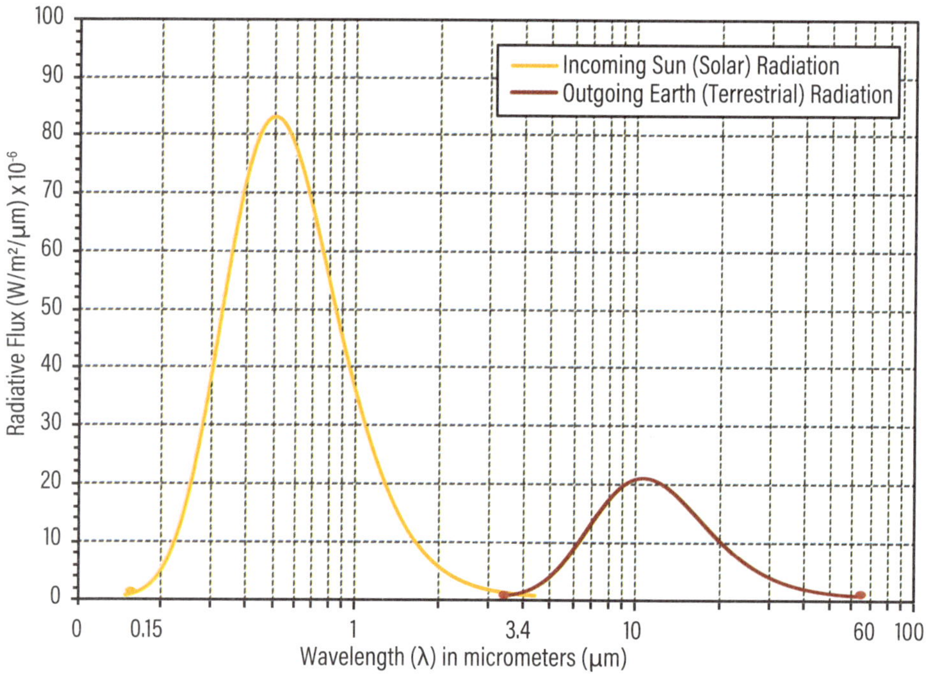

Figure 9-2. Radiant flux of solar and terrestrial radiation.

Of all the GHGs, water vapor is by far the largest contributor to the greenhouse effect. On average, the ratio of water molecules to carbon dioxide molecules in the air is 30,000 to 410. That means that, from a numbers standpoint alone, H_2O vapor molecules contribute 73 times more to the GHE than do CO_2 molecules. On top of that, H_2O vapor is more efficient at absorbing LWIR than any of the other GHGs, and especially CO_2.

The result is that, of all the greenhouse gases, water vapor contributes 95% of the warming GHE to the Earth,[4, 5] as Figure 9-3 shows. CO_2 is a **distant** second place contributor, and the remainder of the major greenhouse gases contributes very little to this protective atmospheric blanket.

Even if CO_2 increased **twofold** from 410 ppm to 820 ppm, or **tenfold** to 4,100 ppm in the future, the story would still be the same. Natural water vapor rules the radiative, heat-retaining capability of the atmosphere. Said simply: Water vapor always dominates the greenhouse effect, never CO_2 nor any other gas in the air, whether natural or human-caused. And so it is, on average, over the entire space and time continuum of the Earth system.

CONTRIBUTORS TO GREENHOUSE EFFECT		
Air Constituent	Symbol	Greenhouse Contribution (%)
Water Vapor	H_2O	95%
Carbon Dioxide	CO_2	3.6%
Methane	CH_4	0.4%
Tropospheric Ozone	O_3	0.01%
Nitrous Oxide	N_2O	0.96%
Other GHG Trace Gases	CFCs & VOCs	0.03%
	Total of all gases:	**100.0%**

Figure 9-3. Various contributors to the greenhouse effect.

[There is more about the spectral properties of greenhouse gases in **Appendix E**]

More on Terrestrial Radiation

How much energy do greenhouse gases actually catch and release? To answer that, we have to delve into terrestrial radiation and the five atmospheric transport mechanisms that move energy back and forth between the Earth's surface and the greenhouse gases. Below are the transport mechanisms mentioned in Chapter 8. They are ordered according to the impact they have on the energy budget.

Radiation: Is the movement of electromagnetic waves of energy.

Convection: Is the upward mechanical motion of molecular bundles of energy.

Latent Heat: Is the heat associated with a phase change that doesn't change the temperature. For example, latent heat of evaporation is the additional heat required to change water (H_2O) from its liquid to vapor state, at a specific temperature. Similarly, latent heat of condensation is the heat subtracted from (i.e., released by) water vapor that changes it from its vapor to liquid state.

Conduction: Is the transfer of molecular vibration energy by liquids or solids.

Feedback: The part of energy that's returned to a source (e.g., the Earth) to either boost or reduce the energy.

Radiation is the mechanism that transports heat from a source of energy to an object by way of propagation of electromagnetic waves. We know that the sun transfers heat to Earth and other planets this way. Another example is the radiation of heat upward from the campfire toward the hands of the camper.

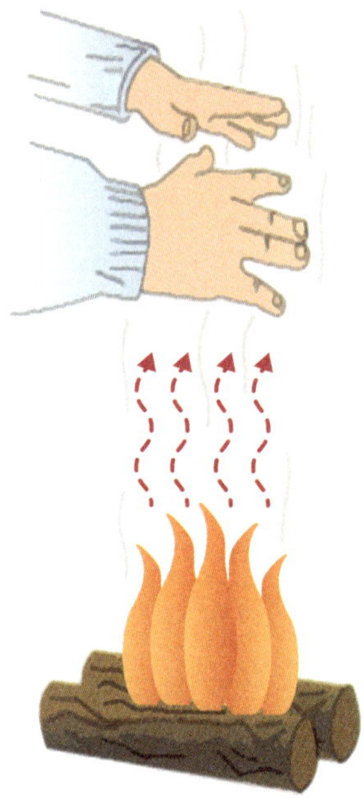

Figure 9-4. Example of radiation.

As the waves from the fire get to the person's hands in the diagram, the hands absorb the energy from the campfire by interacting with the molecules of skin, blood, etc., to raise the temperature, or warmth, of the hands.

Our Earth's land and water, when facing the sun, absorb 51% of the visible radiation coming from it, and the land and oceans warm as a result. At night, the Earth's land and water surfaces "cool" as the Earth rotates away from the sun. Likewise, the hands and bodies of those around the campfire also cool as the campfire dies out. Cooling means that the internal energy and temperature of the object goes down as heat is released or emitted from the object. This is the transport mechanism for the release of energy known as **radiation.**

Convection is the larger-scale physical movement of groups of air molecules, either liquid or gaseous. In Figure 9–5, the air nearest the campfire heats up, raising the internal energy of that bundle of air, and thus the movement (kinetic energy) of that same air bundle, which determines its temperature.

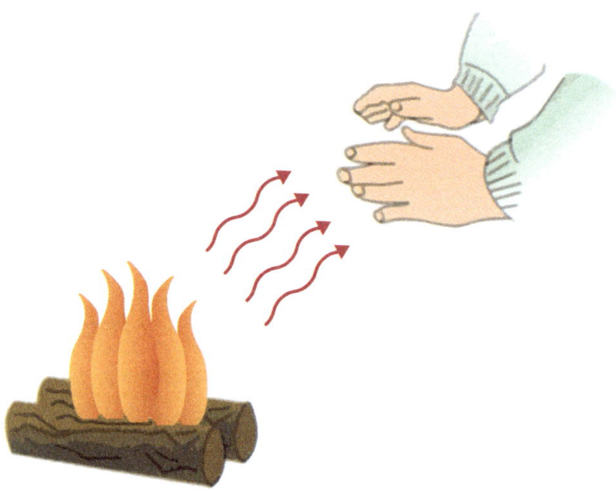

Figure 9-5. Example of convection.

The now more energetic air molecules expand their volume, which is to say their density decreases, relative to their non-heated neighbors. The ensuing action is that this warmer, less dense air rises into the air, relative to its cooler neighbors in the air. A classic example of convection is a hot-air balloon.

For another example of convection, consider a hot, cloudy, summer afternoon on the Atlantic Ocean. Sunlight, peering through a small opening in the clouds, irradiates a relatively small area of the ocean's surface that is exposed to more sunlight than its cloudier surroundings. That sunlit ocean surface area, and the air just above it, gains solar energy over time, making it warmer than the surrounding air and water—forming a hot spot so to speak. The warm air in this hot spot rises and is replaced by cooler air flowing in from the sides. The just arrived cooler air also warms in the sunlight, then also rises, and is replaced below by more cool air. The cycle continues and what we now have is a current of warm rising air and cooler air flowing inward to replace it. That's the convective transport process.

Latent heat is another energy transport process that happens in conjunction with convection. Water molecules at the surface of the ocean absorb energy from the sun, At some point, the H_2O molecules **absorb** enough solar energy to change their state from liquid to gas. We call this change of state the molecules' **latent, or "dormant," heat of evaporation.** This is because the heat stays with, or is dormant in, the molecules as they convectively rise in the atmosphere until they cool sufficiently to change back to the liquid phase. At that point, the gaseous H_2O molecules release their additional heat to the surrounding atmosphere as **latent heat of condensation**, and become liquid molecules again. This is an example of how a rapid and observable convective process of rising humid air, coupled with latent heat of condensation of the water vapor within it, becomes a sustained current. It is an important component of the tropical hurricane formation process. (See Chapter 11 for more details.) Needless to say, convection is an important part of the energy balance equation.

The convective and latent heat transfer processes happen on land surfaces in the same manner as water evaporates from the surface of bodies of water to the atmosphere. Similarly, solar energy assists the movement of water into plants and trees through roots and upward within the plants and trees. A fraction of that water feeds the plants and the rest ultimately exits as water vapor through pores, or stomatal apertures, in leaves, stems, etc. The size of stomatal apertures, and therefore the amount of water vapor that is able to move from leaves to the atmosphere, is variable based on the type of plant and tree. Stomata also serve as input apertures for air used in the photosynthesis process, as well as output apertures for oxygen, the other essential byproduct of photosynthesis. (See Figure 8–7 of Chapter 8.) In other words, plants and trees breathe just like humans and animals do.

Conduction is simply the spontaneous transfer of heat within solid or fluid objects, as exemplified by the iron rod in Figure 9–6.

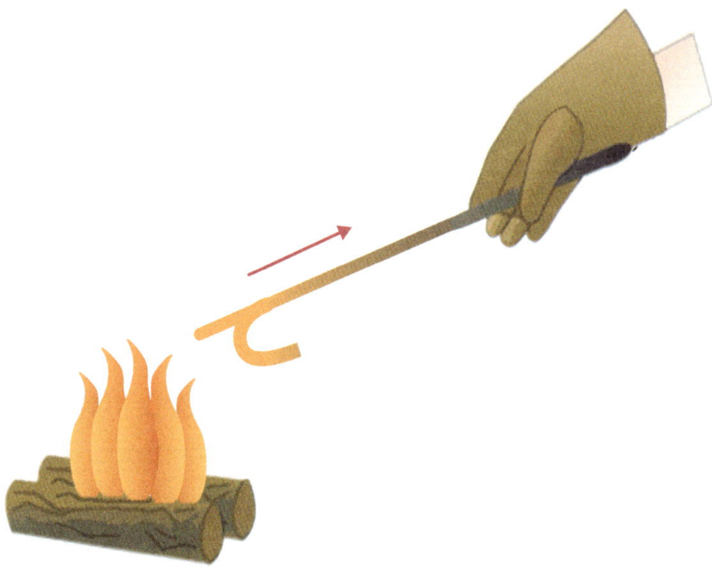

Figure 9-6. Example of conduction.

Think about when you grab a bottle of water or a can of soda out of the refrigerator. Your 98 °F fingers sense the colder 40 °F bottle or can you are holding. The bottle or can is cooling down your fingers, and your fingers are warming up the bottle or can, and its contents, albeit slightly. Here radiation is not involved in the transport process. Heat merely "flows" into one object and out of the other, in order to **balance** out the temperature of both in the process. As the temperature goes up within the molecules of one object, it goes down in neighboring molecules.

An ice cube in a glass of tap water is an example of conduction and latent heat. The liquid tap water cools as it loses energy—that is, temperature—to the ice cube. The ice cube warms and melts as it gains energy from the tap water. This continues until the water, and now melted ice, reach the same temperature. This is called **equilibrium.** Some conduction happens within land and water of the Earth's surface, but it is miniscule in comparison to radiation and convection.

Feedback is the final energy transport mechanism. I saved it for last because atmospheric feedback is a current topic of especially intense disagreement among scientists. It pertains to a number of atmospheric phenomena that can either add to (positive feedback) or subtract from (negative feedback) our global average temperature.

The Cereal Bowl Dichotomy: Feedback Explained

To better understand the difference between positive and negative feedback, consider the simple analogy of an empty, upright, cereal bowl and a marble on a flat tabletop. If you place the marble in the bottom of the cereal bowl and give it a flick, the marble moves up the side of the cereal bowl a bit. Soon gravity takes over and the marble reverses direction, goes down to the bottom of the bowl, up the other side, and back down toward the center again. The marble goes back and forth in the cereal bowl, gradually losing height with each trip due to friction, until it finally comes to rest back at its original position at the bottom center of the cereal bowl. Gravity, friction, and the sides of the bowl prevent the marble from escaping and traveling away indefinitely. This is "negative feedback."

Now, let's do another experiment with the cereal bowl and marble. This time, put the cereal bowl upside-down on the tabletop with the marble on top of the now inverted cereal bowl. Give the marble a flick and this time it rolls off the top, down the side of the cereal bowl, and zooms away until the marble falls off the table or stops due to friction. If that marble gained enough velocity traveling down the side of the cereal bowl, it could go a very long distance before coming to rest. That's because this time there are no boundaries on the marble, and gravity helps the marble escape from the unstable equilibrium atop the inverted cereal bowl. The lack of constraining sides of the cereal bowl this time lets gravity help the marble travel away indefinitely. This is "positive feedback."

The radiative greenhouse effect is probably the most obvious example of a positive (cereal bowl) feedback process.

During daylight, the sun radiates energy into the surface area of the Earth facing the Sun. That particular surface area accepts the incoming solar energy by "catching", or absorbing, it and warming as a result.

The Earth is spinning about its axis. As it does so, the previously mentioned surface area rotates away from the sun and into darkness. Darkness means no more solar radiation to keep that particular surface area warm. Now it cools by "releasing", or re-emitting, terrestrial LWIR up into the atmosphere.

As darkness continues, the greenhouse gases in the atmosphere respond to the approaching terrestrial LWIR by catching (absorbing) "some" of it, and passing the rest of it upward to the outer layers of the atmosphere and ultimately to outer space.

The portion absorbed by the GHGs then re-emits that energy in all directions, some of which goes back, or "feeds back", toward the Earth's surface.

The four-step process repeats continuously, returning to Earth some terrestrial LWIR radiation that would otherwise escape to outer space. This is not additional warming. Rather, it simply prevents some cooling, which would not have been the case without the blanketing effect of the GHGs. This is a subtle, but important, point. The greenhouse gases do not generate new energy to bathe the Earth and increase temperature. The greenhouse gases, and water vapor in particular, are really good at simply intercepting and redirecting, or "feeding back," existing terrestrial energy back and forth and back and forth.

You might think that the greenhouse gases trap more and more terrestrial energy, with the result that more and more heat is trapped, kind of like the run-away marble zooming off the inverted cereal bowl. But they don't.

The reality is that there are at least two other feedback processes also in play, one negative and one positive. The positive feedback is water evaporating from land or sea. Moist air rises via convection, then cools, forming cloud droplets. In that process, as the water vapor changes to cloud droplets, the water vapor also releases latent heat of condensation to the surrounding air. The now cooler cloud droplets are then able to reflect more incoming solar energy back to outer space, an offsetting negative feedback. Harking back to the cereal bowl analogy, what we have is a right side up cereal bowl with temperature rolling back and forth from warming to cooling without getting out of control in either direction.

These are examples of two of many feedback processes. The combined effect of all feedbacks is that they offset each other and translate to the comfortable, average, global temperature of 59 °F that we currently experience.[6] This is in contrast to the freezing −1 °F that would otherwise be the case without these feedback mechanisms. The fact is, the feedbacks help balance global temperature. So how can runaway, catastrophic warming be happening, as climate activists insist?

Who's Driving the Train—Humans or Nature?

We live in an interglacial warming period right now. Our current average global temperature of 59 °F is not constant. But, it is controlled and is **not** in a runaway catastrophic mode. In reality, it has risen slowly over the last 100 years or so, from 57 °F a century ago, to today's value of 59 °F. However, that also begs the question: What is causing this slow rise in temperature?

Those who believe that the Earth-atmosphere system is warming catastrophically argue, without any empirical evidence, that increasing fossil fuel consumption causes increasing CO_2 concentration in the air.[7, 8, 9, 10] This, it is believed, causes more positive than negative greenhouse feedback—an energy imbalance that causes catastrophically increasing temperature. Skeptical scientists, like myself, disagree, and here is why. Figure 9–7 shows Earth's temperature history for the last 500 million years.[11] The white bar along the right hand edge of the plot contains Earth's temperature range during the last 500 thousand years.[12] That range includes five ice ages and five interglacial warm periods (see Chapter 6, Figure 6–7). The red dot signifies our current temperature in the most recent interglacial warm period.

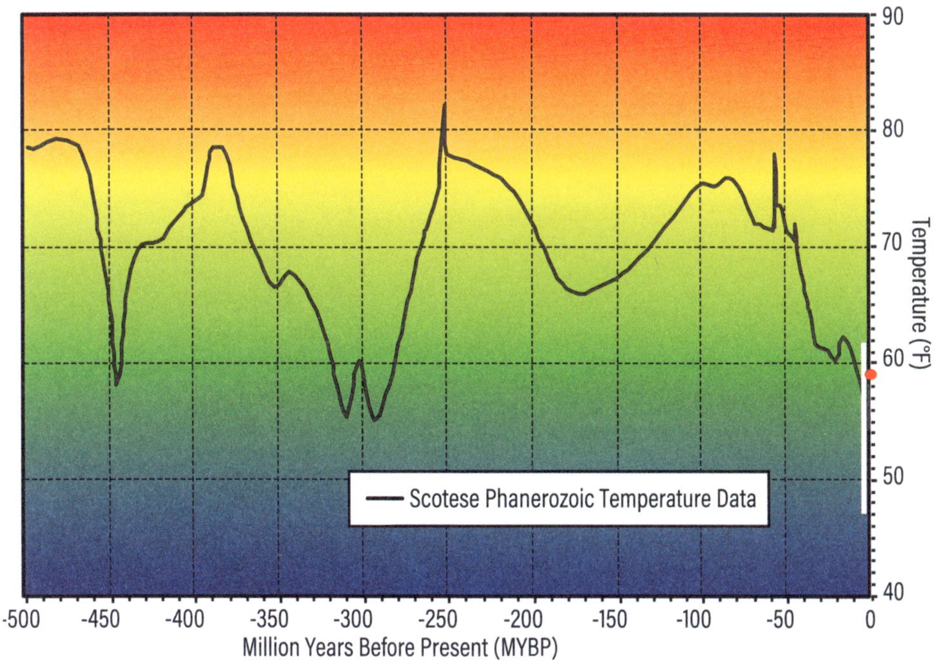

Figure 9–7. Average global temperature during the current Phanerozoic geologic eon.

This evidence demonstrates that the Earth is not burning up now, nor has it done so at any time in the last 500 thousand or even 500 million years. Our planet's temperature continues to vary cyclically in a controlled, stable manner. In fact, we are closer to freezing than burning. If CO_2 and/or humans are the control knob on our climate, it is certainly not obvious from the history. Nor does this evidence suggest any impending disaster.

To conclude this chapter, let me add one more piece to our "controlled climate puzzle"—Figure 9–8, which is an overlay of both temperature and CO_2 histories for the last 500+ million years.[13]

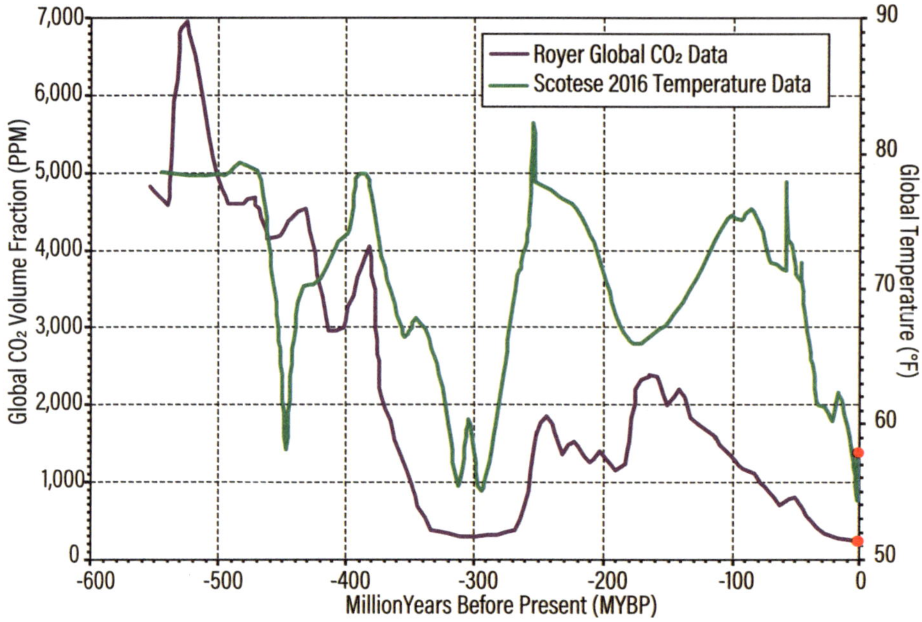

Figure 9–8. Historic temperature levels (green) and CO_2 levels (purple).

This plot looks like it came from a spaghetti bowl. But, if you stare at it for a moment, there are several interesting things to be learned from it, namely:

From roughly 450 to 550 million years ago, both temperature (in green) and CO_2 (in purple) were much, much higher than they are today (at the right hand corner of the plot).

Both global temperature and CO_2 levels are near their lowest levels today (at the lower right hand corner of the plot). The last time this happened was around 300 million years ago.

From about 420 to 450 million years ago, CO_2 levels were quite high (well above 4,000 ppm), while global temperatures were quite low. So much for the catastrophic global warming due to runaway greenhouse gas theory.

Our Earth has been much warmer consistently in the past, than today. Yet, our Earth-atmosphere system has still progressed quite well.

Earth's temperature profile has been sometimes too hot, sometimes too cold, and sometimes just right for the last 500+ million years. But, in all cases temperatures have been both cyclic and bounded. And, at times, temperature has been both in and out of phase with a stark decline in CO_2 from roughly the earliest 7,000 ppm level to today's 410 ppm level. This 410 ppm level is closer to lowest level ever experienced, not the highest level ever experienced.

But, if our average temperature is apparently rising ever so slowly, what is causing it? The chief answer is H_2O. Again, this should not be surprising since water covers 71% of the Earth's surface and clouds on average cover 67% of Earth's land and sea surfaces. And water is the overwhelming contributor to the greenhouse effect, as discussed in Chapter 8.

Other contributing factors for long term **slowly rising and falling** global temperatures are planetary motion, continental drift (aka plate tectonics), volcanic activity, and changes in solar activity, as shown in Chapters 4 and 5.

It's not clear what else might contribute to the small-scale temperature fluctuations that ride on top of interglacial warming. It hides within the limits of today's thermometers, as well as the spatial, and often sparse, distribution of measuring stations, especially in the cold Arctic and Antarctic regions of the planet,[14] as addressed in Chapter 6.

We just don't know at this point. There are a number of atmospheric scientists, including physics Nobel Laureate Ivar Giaever (1973 Nobel prize in physics), who have concluded that we really don't know if our global surface temperature is rising or falling because whatever it's doing is so negligible.

This difference of opinion came to a head back in January of 2015 when President Barack Obama warned " ... No challenge poses a greater threat to future generations than climate change."[15] Nobel Laureate Giaever responded: "That is a ridiculous statement ... How can he say that? I say this to Obama: Excuse me, Mr. President, but you're wrong, dead wrong ... How can he say that? I think

Obama is a clever person, but he gets bad advice. Global warming is all wet."[16]

Instinct alone should tell us the sky is not falling, the atmosphere is not catching fire, and a state of uncontrolled, runaway warming is unrealistic. We know better, and there is still more historical evidence to come in the remaining chapters.

CHAPTER TAKEAWAYS

1. H_2O vapor dominates the greenhouse effect, accounting for 95% of the effect compared to CO_2's paltry 4%. And neither adds heat to the atmosphere. The greenhouse effect simply captures and redirects, toward Earth, some of the LWIR terrestrial radiation that would otherwise escape to outer space. The result is a hospitable 59 °F average global temperature, rather than a frigid –1 °F temperature without greenhouse gases in the air.

2. **Major** warming and cooling trends (i.e., 10 °F to 15 °F swings) do occur infrequently at the Earth's surface. Minor warming and cooling trends also occur more frequently and are "blips" riding atop the major trends. But all are due to the same complex and cyclic interaction among water vapor in the air and the inertial and gravitational effects of spin, orbit, tilt, precession, and eccentricity.

3. Our Earth has been much warmer in the past than today, and the system has still progressed as designed—always balancing itself out.

4. The bottom line is: Energy is conserved in our air-land-water system. The result is a comfortably warm global environment that is gently rising in its balanced, normal, and cyclic manner to beneficially sustain and nourish all life on this planet.

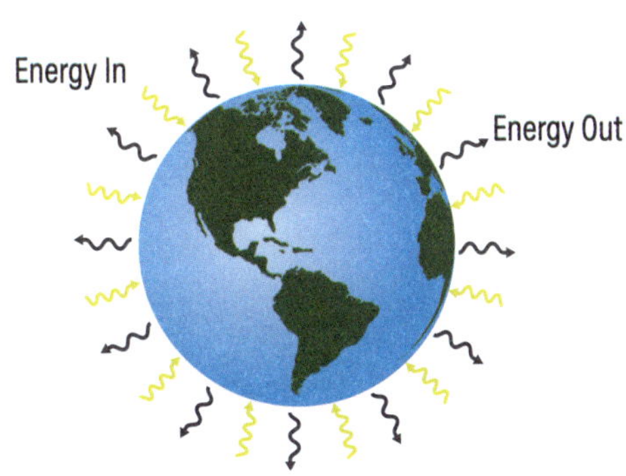

chapter 10

Tornadoes

The Worst Experience of My Life—Chicago, Illinois, April 21, 1967

When I was a youngster, back in the 1950s, people often labeled tornadoes as "the finger of God." Of course, I don't know if that characterization is accurate, and there are some other things I don't know about tornadoes. But one thing is certain: I identify with tornado victims. I experienced one first hand—an extremely violent and deadly EF4 tornado on Friday, April 21, 1967.

On that fateful day, I was just getting off the "L" train on the south side of Chicago, heading home for a relaxing weekend after my classes at DePaul University had ended for the week. The "L" or "EL" is an adored nickname for the commuter trains that run throughout the city of Chicago on "el"evated tracks mounted on raised steel girders that are about 25 feet off the ground.

I thought it unusually warm and humid on that peculiar spring day. The cloudbank above was thick and ominous looking, and the world seemed much darker than it should have been. But most unnerving of all was the lack of any breeze. I sensed something was

ACRONYMS USED IN THIS CHAPTER

Acronym	Definition
L or EL	The "elevated" commuter train in Chicago
RADAR	Radio Detection and Ranging
VFW	Veterans of Foreign War

badly wrong, and, apparently, so did the hundreds of others simultaneously exiting the train onto that narrow elevated platform at 63rd Street and Loomis Avenue. The only order of business at that muggy, stifling moment was to get down to ground level and find cover.

We were packed tightly going down the stairs. I was in the middle of the crowd; unable to hold onto either railing, thus totally helpless if anyone fell on the way down. Luckily, none of us did. It was scary, but, at the same time, a surprisingly calm and orderly movement as we all hustled down the narrow stairway and into a nearby grocery store. There were several minutes of intense rain, hail, and lightning and then the storm departed as quickly as it arrived.

We all then went on our separate ways, relieved that we had dodged a weather bullet. We were the lucky ones. As I would find out later, that severe thunderstorm that descended upon us had minutes before been a violent tornado that devastated a 16-mile long by 500-foot wide swath of the southwestern Chicago suburb of Oak Lawn, Illinois.

When I arrived at my childhood home near Midway airport, I still did not know what had just happened four miles to the south. I dumped my dirty clothes in a pile to wash later, grabbed the keys to the family car, and headed out for an evening of what I thought would be pizza, TV, and fun with a few close friends.

It was getting dark when I left the house, a mere two hours after the encounter with the storm at the "L" train station. As I drove southward, my first indications of trouble were tree branches littering the street and unlit streetlights.

Homes and businesses were all dark as well. It was completely black and eerily silent outside, with my car's headlights now the only source of light in front of me. Then, in the distance, I began to hear the faint sound of sirens. At this point, the tree branches and debris were getting larger and more massive, and broken glass was everywhere. I parked the car and walked the remaining few blocks to my friend's house in complete darkness except for the flashing red lights in the distance ahead of me. I had no flashlight, but I could see that about a half a block in front of me was 95th Street and Southwest Highway, and something was clearly wrong. Yet, it wasn't until I came upon my friend, minutes later, standing motionless in his front yard, that I grasped what had happened when he murmured the single word "tornado."

Emergency vehicles were scurrying back and forth along the street in front of us. Maybe we could help, we thought. And so we carefully approached the exponentially increasing, disastrous, and horrifying scene that was quickly unfolding with each step forward.

There were no words to describe the devastation as we reached 95th Street. Wood- frame homes, a gas station, and a Jewel grocery store were reduced to piles of rubble. Brick homes mostly survived, but many had their roofs ripped off. Portions or complete walls, sucked into the tornado, had been spit back out along the storm's path. Windows were glassless, with mere remnants of blinds and shredded drapes remaining.

Then came the bus terminal and garage, with 15-ton public transit buses tossed like toy models hundreds of feet away. One was planted firmly in the front living room of a home. Cars were contorted and mangled so badly you could not tell their make or model. It was a traumatic sight.

We asked what we could do to help, and someone yelled for us to assist with communications at the nearby Veterans of Foreign War (VFW) post a couple of blocks away. Puzzled at how that would help, we nevertheless immediately headed over there. We walked into the vestibule of the VFW hall where several men were whispering and trying to set up radios in near darkness. Whispering? Then, we glanced into the dance hall, also dimly lit by a couple of light bulbs connected to a small generator out back. On the floor were neat rows of white sheets, some flat on the floor, too many others with crumpled bodies beneath the sheets. It was the makeshift morgue. Even now, some 52 years later, I have not forgotten that humbling image. We helped a bit, and then left, melancholy and changed forever. That violent tornado killed 33 men, women, and children, and injured 500 more in a matter of minutes.

The Tornado Story

This story clearly illustrates that that a tornado is an isolated, short, and life-changing event. But, it is not a normal occurrence or a trend in and of itself. Similarly, any weather event is **purely a snapshot of atmospheric conditions at a single point in time. It's not a trend over a long period of time. It is not climate.**

And as such, it bothers me that it has become all too commonplace to link these intense, but isolated, weather phenomena to "global warming" or "climate change." Activists insist we have more frequent and more powerful tornadoes today than at any time in the past. And they blame it on global warming. Over the next several pages, I shall show you the clear, verified evidence (measurements and observations) to demonstrate that their hypothesis is false. Tornadoes today are not more frequent, nor more powerful.

In keeping with our big picture philosophy and perspective, Figure 10–1 illustrates the spatial distribution of tornadoes, worldwide.[1]

The brown splotches are where tornadoes occur most frequently. A large fraction of the Earth does not experience tornadoes but the United States certainly does. The U.S. leads the way with about 75% of the Earth's known tornadoes measured in terms of frequency (number of tornadoes per year) and intensity (i.e., kinetic energy, measured in terms of the rotational speed of tornadoes). A few other countries of the world are not immune either. However, they tend to have far fewer tornadoes than the U.S. But the ones they do have can be every bit as intense as the tornadoes in the U.S.

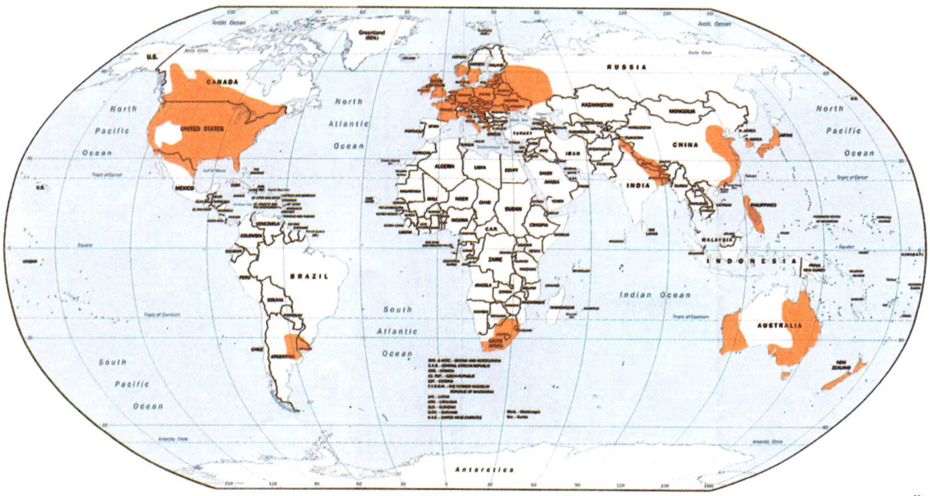

Figure 10-1. Global tornado distribution.

Since the U.S. has such a large fraction of tornadoes, as well as the best records, it's sensible for us to focus on the U.S. as being representative of worldwide tornado events.

What is it that makes the U.S. so prone to tornadic activity? To answer that, we must again look at the big picture, and direct our attention to the region from the North Pole downward through the Midwest portion of the U.S., all the way down to the Mexican border. Especially during the spring, and even well into the summer, cold Arctic air masses are still traveling down from Canada. But different from winter, on occasion, these cold northern air masses now collide with warm, humid U.S. air masses coming from the south and southwest.

The result of such an atmospheric interaction of warm and cold is what we call a strong temperature gradient (a **strong change in temperature** over a very short amount of time and distance). And when the cold air mass happens to ride up and over the warm air mass in the collision process, then the movement of air is even stronger.

Remember that warm air rises and cold air sinks, and the contention between the very cold and very warm air—the result of a strong temperature gradient—is turbulence, which is a significant factor in swirling tornado formation. The strength of temperature gradients are most severe in the spring.

Warmer Arctic, Canadian, and U.S. regions manifest themselves in the form of weaker overall temperature gradients. Therefore, a lesser number (frequency) of tornadoes and/or their associated intensities should result. So, claims that global warming is causing more frequent and severe tornadoes are nonsense. Alarmists are arguing against themselves, because warming means fewer tornadoes. If our planet is warming catastrophically, we could cease to see tornadoes.

Let's take a look at historical tornado data that shows us that these impressive storms have not been getting worse over time. There are two metrics to use to analyze tornadoes. One is to simply count the number of tornadoes that have occurred in the U.S. year-by-year. It's the easiest to do and provides an adequate picture. The other, and I believe better, metric is to look at tornado intensity, or accumulated energy. We'll look at both number and energy.

Tornado Frequency

We'll begin by counting the number of tornadoes that have occurred each year since detailed record keeping officially began in 1950.[2] We refer to these numbers as **tornado frequency** over the last 69 years worth of data (1950–2019). Since U.S. tornadoes are 75% of worldwide tornadoes in terms of frequency and intensity, the U.S. tornado frequency metric in Figure 10-2 is representative of the global tornado metric.

At first look, Figure 10-2 suggests that tornadoes are, in fact, increasing, as climate activists suggest. The blue zigzag line is the actual number of tornadoes that were recorded during the period from the beginning of 1950 through 2019. The black line is the linear trend line over the period.

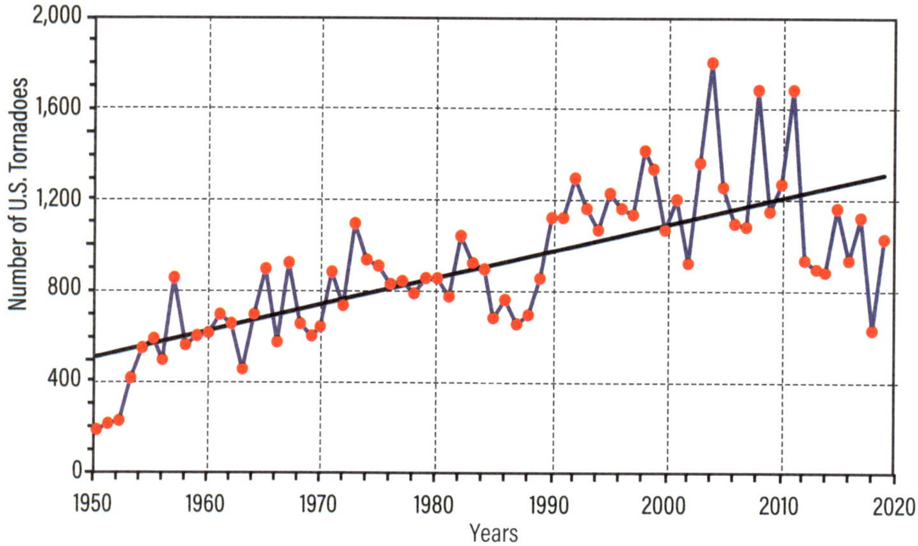

Figure 10-2. Number (frequency) of U.S. tornadoes that occurred between 1950 and 2019.

However, Figure 10–3 is the same data as in Figure 10–2, except that in Figure 10–3 the data is split into two separate trend lines, from 1950 to 1990 on the left, and from 1990 to 2019 on the right.

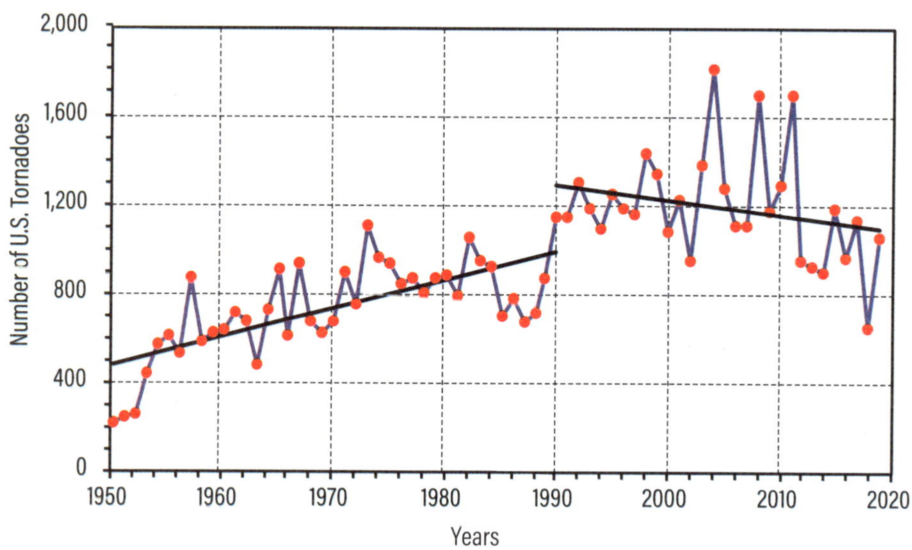

Figure 10-3. Number of U.S. tornadoes that occurred between 1950–1990 and 1990–2019.

Now Figure 10–3 tells a different story than Figure 10–2, namely increasing tornadoes early on, and then a change of direction around 1990, followed by a decreasing trend in the number of tornadoes for the last 29 years since 1990. Why? What changed around 1990? The answer was mostly due to advances in the methods and quality of tornado counting, tracking, reporting, and warning technologies. Over the years between 1950 and 1990, population grew in rural areas. As a result, more and more tornadoes (especially the small ones) were counted as rural populations grew and people noticed and reported tornadoes happening. Severe weather tracking technology also matured during the same period (1950–1990).

In the 1950s, RADAR evolved from detection of large, hard targets such as airplanes to detection of small, softer, weather-related targets such as hail, rain, and snow. The newer RADARs assisted extensively with counting and location of tornado-prone storms. Thus, the result was a steadily increasing number of tornadoes being reported since 1950.

However, another step upward in technology happened in the late 1990s when pulsed Doppler RADAR matured for both weather positioning and motion measuring. This facilitated much better accuracy in the detection, tracking, and scrutiny of rotating vs. non-rotating severe storms. The result has been better ability to detect tornadic rotation, and thus sort out other less-severe, straight-line storms that would have been counted as tornadoes in the past. Technology is one important reason for the declining trend in tornadoes since the 1990s.

Tornado Intensity

Next, we'll look at tornado **intensity (energy)** metrics to see if this method agrees with tornado **frequency** metrics.

Tornado intensity is categorized on a scale of 0 through 5, as shown in Figure 10–4.

2007 ENHANCED FUJITA (EF) TORNADO INTENSITY SCALE		
Category	Wind Speed (mph)	Force Descriptor
0	65–85	Gale
1	86–110	Weak
2	111–135	Strong
3	136–165	Severe
4	166–200	Devestating
5	> 200	Incredible

Figure 10-4. The 2007 Enhanced Fujita "EF" tornado intensity scale.

The estimated intensity is based on observed wind speed and damage, and is named after Dr. Ted Fujita, a meteorologist who developed the scale originally in 1971 and then updated (enhanced) it in 2007.[3]

The starting point for a tornado energy metric is the same as it was for a frequency metric, namely the number of tornadoes of all types over the reporting period of 1950 through 2019.

Here is Figure 10–3 again, except now retitled as *"Figure 10–5. All U.S. Tornadoes (EF0-EF5)"* for reference.

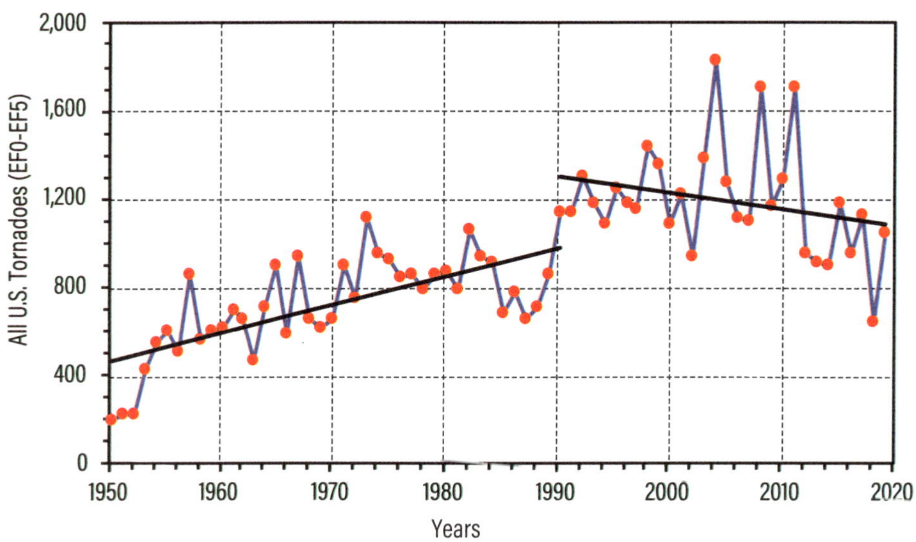

Figure 10-5. All U.S. tornadoes (EF0-EF5).

This graph represents the total count of 63,802 tornadoes of all intensities (EF0 through EF5) reported from January 1, 1950 through November 30, 2019.

Now, we're going to look at the number of U.S. tornadoes broken down by each of the EF tornado intensity (aka severity) categories that were described in Figure 10–4. This will give us a picture of how tornado severity has varied over the last 69 years.

Note that I keep the x-axis (time scale in years) constant at 1950–2020 throughout the this chapter. However, I will shrink the y-axis scale in each successive plot from Figure 10–6 through Figure 10–14 so that you can clearly see the magnitude variability of tornadoes from year to year.

Figure 10–6 and 10–7, below, are the EF0 and EF1 tornadoes—the weakest subset of tornadoes.[4] They constitute 80% of all U.S. tornadoes, cause the least amount of deaths, injuries, and damage, and generally have trended upward over the last 69 years.

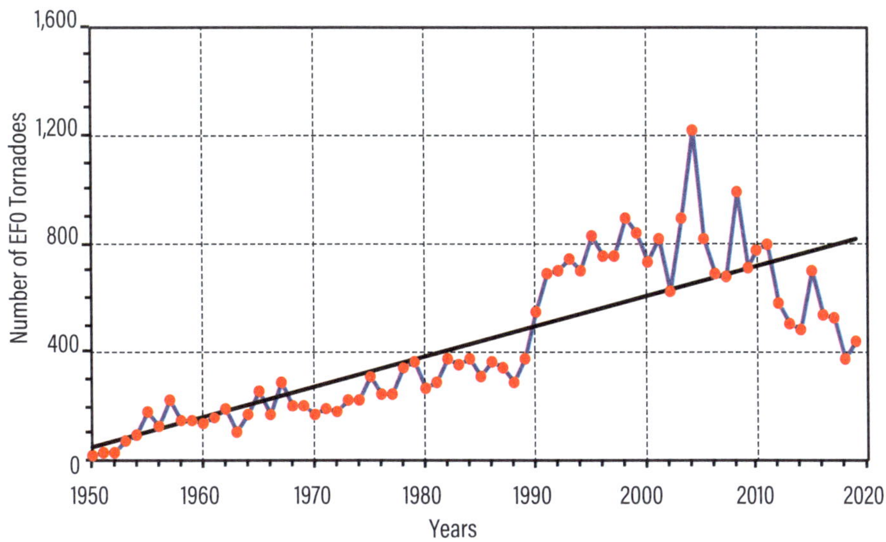

Figure 10-6. U.S. EF0 tornadoes (1950-2019).

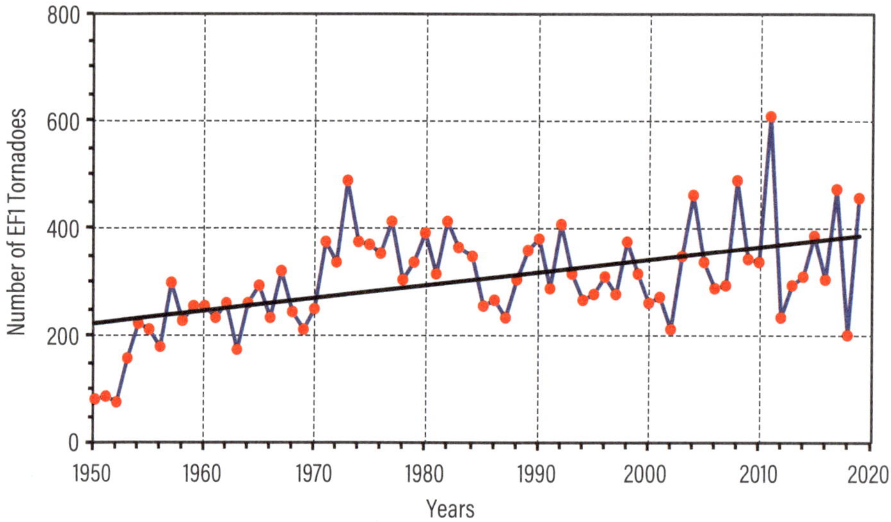

Figure 10-7. U.S. EF1 tornadoes (1950-2019).

On the other hand, Figures 10–8 through 10–11 are the strong (EF2), severe (EF3), devastating (EF4), and incredibly violent (EF5) storms that are much less likely to be missed or unreported, due to their impact on society. As you will see, these intense tornadoes all have the following notable characteristics—fewer numbers in each succeeding category and a decreasing trend since 1950.

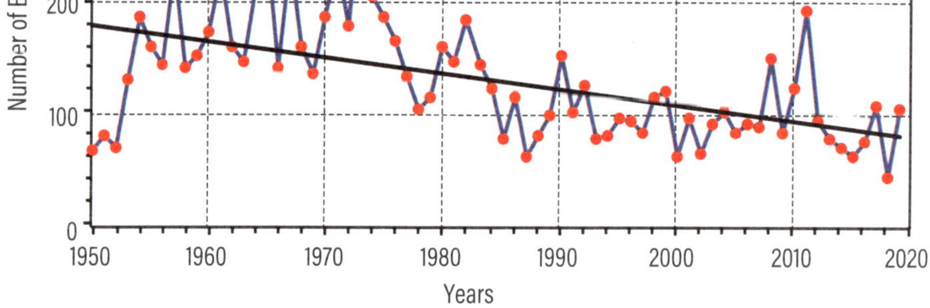

Figure 10-8. U.S. EF2 tornadoes (1950-2019).

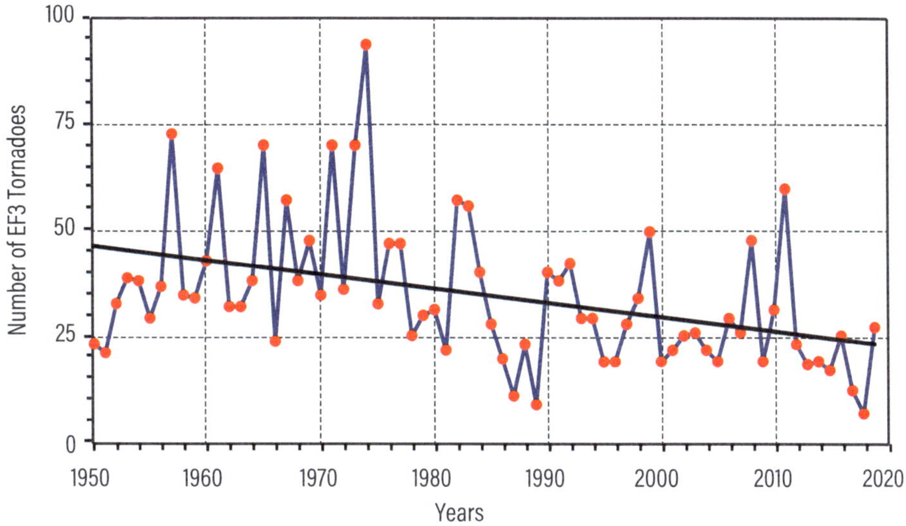

Figure 10-9. U.S. EF3 tornadoes (1950–2019).

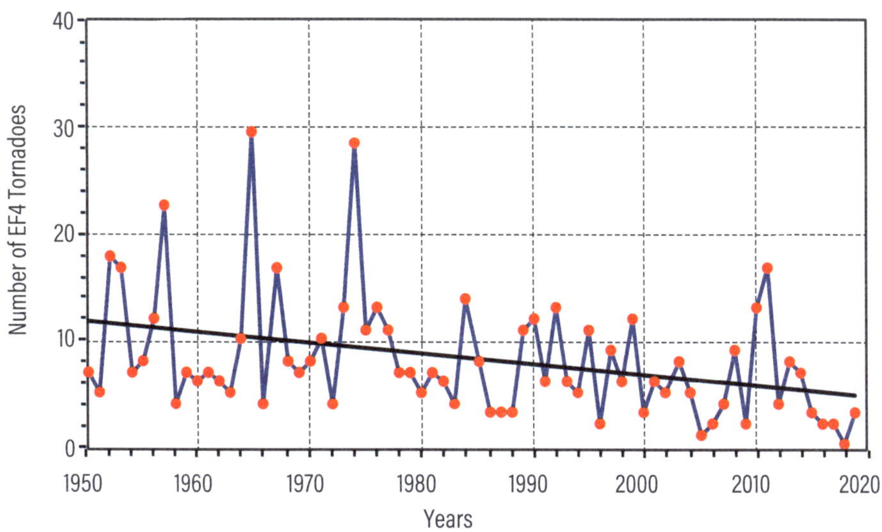

Figure 10-10. U.S. EF4 tornadoes (1950–2019).

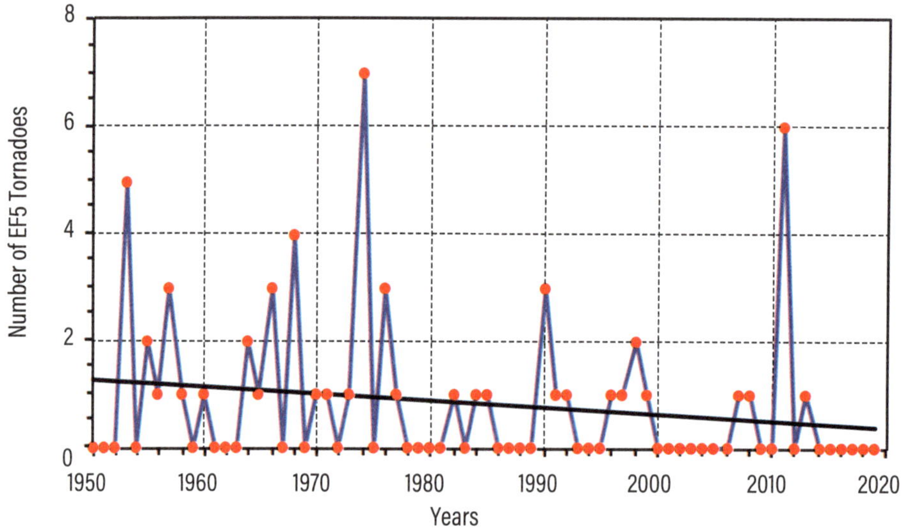

Figure 10-11. U.S. EF5 tornadoes (1950-2019).

It is a common misconception that tornadoes are increasing and becoming more violent. As is clear from the preceding plots, the tornadoes that we fear most have not increased in number or intensity over the last 69 years of reporting.

Tornado-related Deaths, Injuries, and Property Damage

Figures 10–12, 10–13, and 10–14 show U.S. death, injury, and damage statistics during the period 1950 through 2018. [Note: Tornado death, injury, and damage numbers are not yet finalized for 2019.]

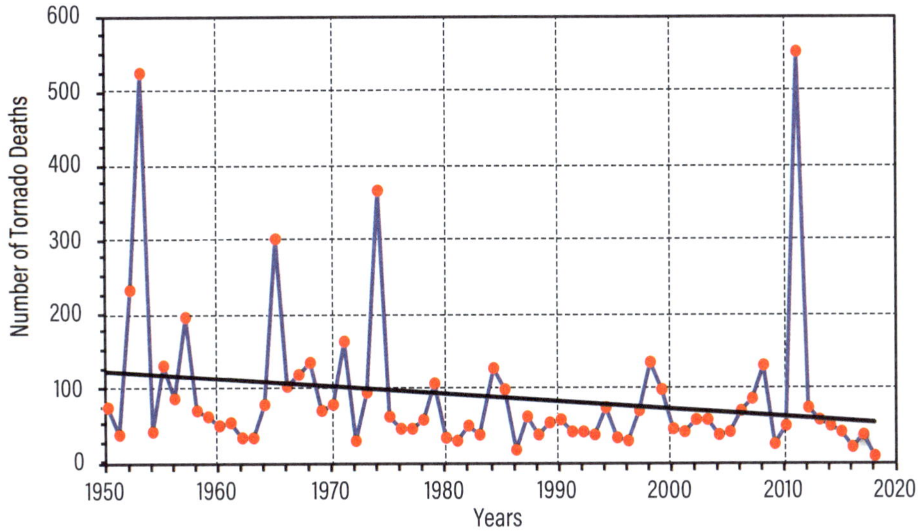

Figure 10-12. All U.S. tornado deaths.

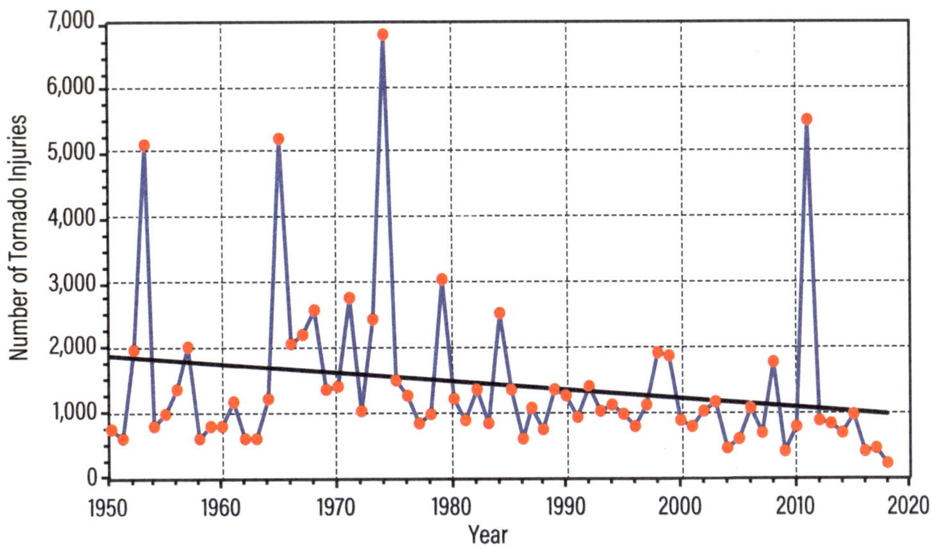

Figure 10-13. All U.S. tornado injuries.

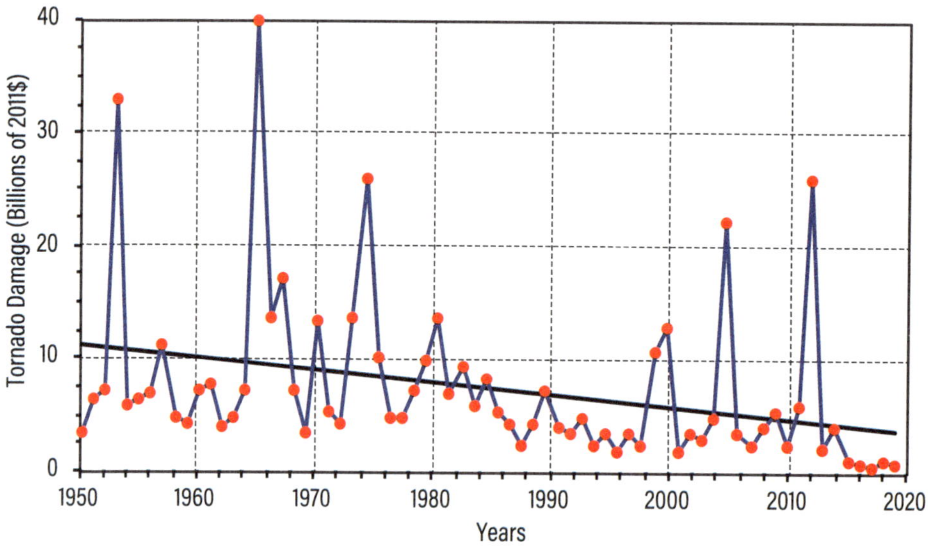

Figure 10-14. All U.S. tornado damage.

Deaths, injuries, and damage to property all have trended downward over the long haul since 1950, as indicated by the black trend lines, above. However, there are some notable exceptions that are common among the death, injury, and damage plots. Those exceptions are 1953, 1965, 1974, and 2011. In those years, tornado deaths, injuries, and damages all were noticeably higher than the average for the 69-year period. Also note that these four years of spikes in death, injury, and damage also correlate well with spikes of EF4 and EF5 tornadoes in the same years.

Common sense tells us that death, injury, and damage correlate with tornado severity. But, what the plots do not portray is that death, damage, and injury from tornadoes are also dependent on geography—where the tornadoes occur. For example, if a violent tornado hits in a large urban area like Chicago, the resulting death, injury, and damage toll is, inevitably, a whole lot higher than would be the case for a violent tornado that hits a rural cornfield.

CHAPTER TAKEAWAYS

1. Global warming should decrease, not increase, the intensity of tornadoes due to relaxed spatial temperature gradients associated with warming as springtime warm and cold air masses collide.

2. Since 1950, tornado intensity, deaths, injuries, and property damage have all trended downward. Tornado frequency has also trended downward, except for weak tornadoes, which have trended upward due to technology and reporting advances.

3. Weak EF0 and EF1 tornadoes have fewer injuries, deaths, and damage associated with them, relative to the EF2-EF5 severe tornadoes. Hence, even given the preponderance of the EF0-EF1 tornado count, it is the EF2-EF5 tornadoes that cause the majority of injuries, death, and damage.

4. Tornado destruction is also dependent on the geography where the tornado occurs.

5. It is incorrect to generalize that **all** tornadoes are increasing. The severe EF2-EF5 tornadoes are not increasing, nor are the injuries, deaths, and damage associated with them.

chapter 11

Tropical Hurricanes, Typhoons, and Cyclones

Hurricane Hunting—1969

After graduating from DePaul University with a Bachelor of Science degree in physics in 1968, I switched interests and began graduate school in the Meteorology Department at Florida State University that fall. Working as a research assistant, my focus was on physical and marine meteorology in particular. In the summer of 1969, I took a full-time position as a meteorologist with the Navy Weather Research Facility at the Norfolk, Virginia Naval Air Station.

My work there focused on Atlantic hurricanes (aka typhoons in the Pacific Ocean), and specifically on a hurricane modification project called "Stormfury."[1] The goal of the 1969 Stormfury experiment was to seed a fully developed hurricane with silver iodide and measure its effect on the storm.

The hypothesis for this experiment was that injecting silver iodide particles into hurricane eye wall clouds from above would transform super-cooled water droplets to ice crystals, thereby releasing additional latent heat of fusion energy into the hurricane. This energy transformation would redistribute the overall energy of the storm

ACRONYMS USED IN THIS CHAPTER

Acronym	Definition
ACE	Accumulated Cyclone Energy
IBTrACS	International Best Track Archive for Climate Stewardship
JWTC	Joint Typhoon Warning Center

Acronym	Definition
MSL	Mean Sea Level
RADAR	Radio Detection and Ranging
TC	Tropical Cyclone

UNITS OF MEASURE USED IN THIS CHAPTER

Base Quantity	Unit of Measure	Symbol
Pressure	millibars	mb
Speed (Maritime)	knots	kt or kts
Temperature	degrees Celsius or Fahrenheit	°C or °F

over a larger area, reduce wind speeds, and possibly even act like a steering wheel to turn the storm's direction.

Plans were in place, and all we needed was a good hurricane to test the hypothesis. Primarily for safety reasons, it had to be a well-established hurricane that was far enough out in the Atlantic Ocean and tracking north so as to not pose a threat of landfall within the ensuing 24 hours. Consequently, the Project Stormfury team waited patiently for several years for just that "right" hurricane to seed, plus presidential approval to execute the experiment.

The "right" storm, and subsequent approval, finally happened in early August of 1969. And, as it turned out, this would be the one and only time a hurricane has ever been seeded. Incredibly, I was able to be a member of the team that would fly into the eye of Hurricane Debbie to record of the results of the seeding effort.

The U.S. Navy's Weather Reconnaissance Squadron (WEARECONRON) Four, or VW-4—the Navy's famous "Hurricane Hunters"—executed the test on Hurricane Debbie on August 18th and 20th of 1969. They were the Navy's elite unit whose mission since 1945 was to track the position and strength of all North Atlantic storms. The unit consisted of a team of incredibly talented pilots and maintenance personnel who flew a half-dozen or so Lockheed WC–121N aircraft in and out of Atlantic and Caribbean storms.

The WC–121N hurricane-hunter aircraft were modified versions of the Lockheed "Super Constellation" passenger aircraft that Trans World Airlines (TWA) used to fly back in the 1960s.

Figure 11–1 is an August 1969 photo of one of the hurricane-hunter aircraft on the tarmac at the Naval Air Station at Jacksonville, Florida.

Figure 11-1. U.S. Navy WC-121N "hurricane hunter" aircraft. Photo by Bill Pekny

For safety reasons, the Navy utilized these propeller-driven, reciprocating-engine aircraft instead of more modern jet-powered aircraft. Jet engines were not sufficiently reliable in the high liquid water content levels of hurricanes. Hurricane hunter modifications to the traditional "Super Constellation" passenger plane configuration include extended range fuel tanks on the wingtips of the aircraft, as well as vertical and horizontal-scanning microwave RADARs above and below the fuselage at the middle of the aircraft.

Figure 11–2 is a close-up of the two RADARs.

The tall dome on top of the aircraft houses the short range, vertically scanning RADAR. The wider, but flatter, dome on the underside of the aircraft houses a long range horizontally scanning RADAR. Together they provided both position tracking and vertical profiling information during hurricanes.

Figure 11-2. Horizontally and vertically scanning RADAR domes on U.S. Navy WC-121N "hurricane hunter" aircraft. Photo by Bill Pekny

The Stormfury hurricane-hunting flights last about 12 hours in the air, and Hurricane Debbie was no different. Flights during seeding operations consisted of about 2–3 hours of travel out to the hurricane, 6–7 hours flying at an assigned pattern and altitude within the storm (Figure 11–3),[2] and then a 2–3 hour flight back to home base.

As illustrated in Figure 11–3, once we got near Hurricane Debbie's location in the middle of the North Atlantic Ocean, the pilots entered the storm's outer rain bands perpendicular to the direction of movement of the hurricane as a whole. Next, we proceeded directly through the eyewall and into the eye of Hurricane Debbie and exited out the tail end of the hurricane. Then we repeated the process.

The "interesting" part of the flight pertained to the low altitude of our inflow-monitoring pattern. Flying at a mere 1,000 feet above the surface of a normal ocean for hours on end was intense in its own right. It was even more intense when flying above a turbulent, churning, white foam, ocean surface in the middle of a full-blown hurricane.

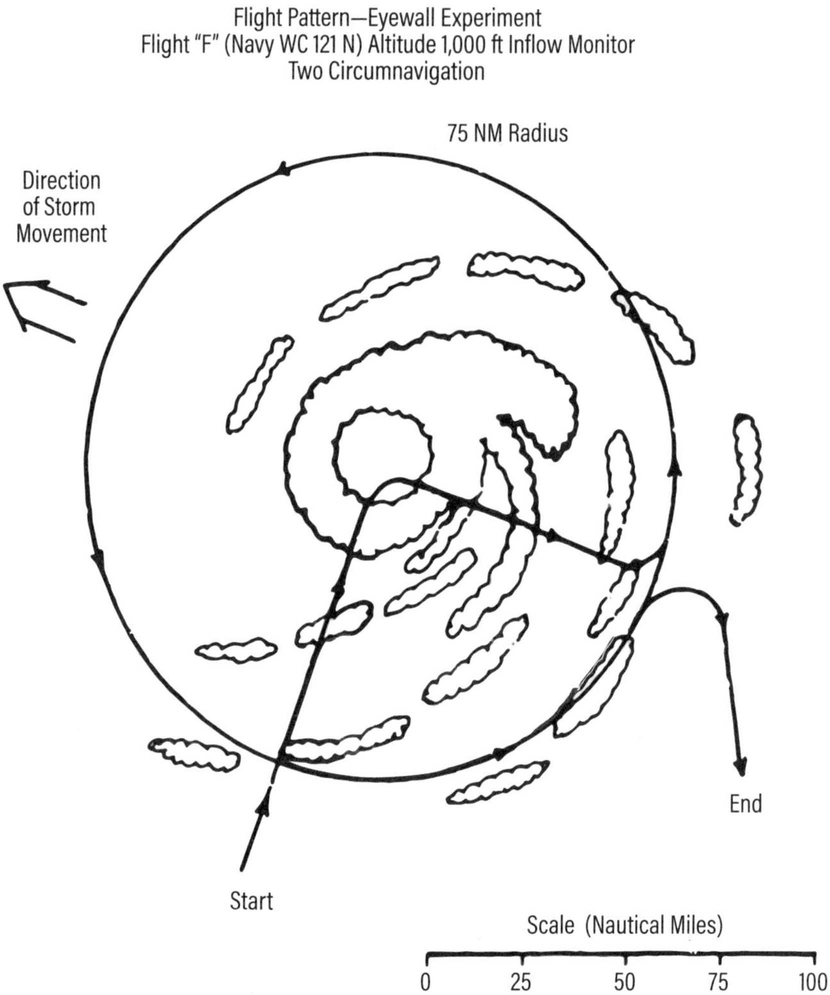

Figure 11-3. U.S. Navy Hurricane Hunter low-level inflow flight pattern.

Near the top of the storm (~40,000 feet high), U.S. Air Force Hurricane Hunters monitored the goings on at that level, and Navy A4 Intruder jets dropped silver iodide into the eye wall from an altitude of 7,000 ft.

As I recall, our crew consisted of about 20 people. In the cockpit there were the pilot, co-pilot, and flight engineer. Nearby, in the forward portion of the aircraft, were at least two backup pilots. All of them were well-trained and experienced Navy officers.

The "aft" (rear) section of the aircraft included more Navy personnel. They served as navigator, radio operator, combat information center officer, and doubles for each of them. Meteorologists such as myself, other technicians, and a **cook** rounded out the remainder of the flight crew. We typically rotated between assigned duties and rest periods about once an hour in order to stay sharp, especially during periods of high turbulence.

Surprising to me, the cast of crewmembers also included a cook. I don't know where they got those guys, but they were wonderful cooks. Each plane had one, and on the way home after a hurricane mission, they prepared fantastic hot meals. I guess the Navy must have figured that anybody devoted enough to fly into hurricanes at least deserved a good meal along with hazardous duty pay.

My specific job, as a crewmember and RADAR meteorologist, was to record meteorological events before, during, and after the seeding of Hurricane Debbie. I sat in front of a RADAR scope with a camera bolted to the front of it, along with a clock and aircraft altitude indicator light-piped into the screen image before me. My job was relatively simple: sit there and monitor the images from the vertically and horizontally scanning weather RADARs for significant meteorological events. I say "relatively simple" somewhat with tongue-in-cheek. Taking pictures during the more turbulent episodes, like when cutting through an eye wall or rain bands, were truly "white knuckle" moments at first. But eventually, the turbulence became routine for this rookie hurricane hunter.

One more unique thing about flying at a 1,000-foot altitude in a hurricane was the environment inside the aircraft. Airplanes, in general, didn't have bulky and heavy air conditioning units on board. They simply pumped in **normally** cool outside air that was then heated as required to maintain a comfortable level. There was no such luck when flying at 1,000 feet in the middle of a hurricane. Also, all of the on-board, tube-based electronics of the time added more heat for us to contend with. Finally, each crewmember had to wear fireproof flight suits and gloves, as well as a "Mae West" life preserver during the flights. Needless to say, it was hot and damp inside the hurricane hunter aircraft.

By far, the most amazing things about those flights were the airplanes themselves, the pilots, and the ground maintenance crews that kept the aircraft in tip-top shape, so the pilots, with their unbelievable skills, could safely fly them. The planes were fairly old by 1969, but performed flawlessly in spite of the flight conditions. There were not many windows in the plane, except in the break area. During rest periods, I would look out the window and marvel at the flexibility in the wings of that aircraft. They flapped slowly, like a graceful bird in flight.

And then there were those magnificent pilots. Beside my RADAR screen was an altimeter so I was always aware of the aircraft's height above the ocean. Even during the turbulent periods, flying through rain bands and eye walls, the pilots always kept the aircraft really close to their assigned altitude of 1,000 ft. They were so well trained that they could feel looming updrafts and downdrafts and react almost instantaneously to maintain the proper altitude.

Their other unusual flight prerequisite was to keep the wings as level as possible, even while turning, in order to keep the RADARs working properly. The pilots also managed to do that well, since our RADARs did not fail. "Skidding" a plane through a turn was uncommon for both plane and pilot, since a normal turning maneuver would have been a combination of banking and turning. This was not the case in a hurricane hunter aircraft.

Yet as hard, or stressful, as flying into a hurricane may have been, the most common refrain was that the most challenging part of the entire flight was taking off. This was due to the tremendous load of fuel needed for the lengthy missions.

The Project Stormfury team successfully executed the seeding of a full-blown hurricane and the gathering of data. However, post-test analysis of the effort was inconclusive. The energy did appear to spread out, and the hurricane did turn further northward. But the team could not determine whether the seeding caused the energy redistribution and the turning motion, or if the hurricane would have done the same thing naturally.[3]

Nevertheless, I've had a keen interest in hurricanes ever since. Today, climate activists are claiming that global warming is causing more and increasingly severe hurricanes. That doesn't jive with my experience on many different levels. To them I simply say: "Show me your global, climate-scale evidence that supports your claim." They can't, because it doesn't exist. Furthermore, more heat does not simply translate to either more or stronger storms. **A hurricane is an intense weather event. It is not a climate event.**

However, I am in no way diminishing the importance or severity of any of these isolated extreme weather events. They are all bad news, and I have the deepest sympathy for any and all of those who have lost peace, property or loved ones to any violent weather event of any kind.

That being said, in the remainder of the chapter, I'll challenge the climate activist claim that global warming causes more and increasingly severe global tropical cyclones. I'll do this with cold, hard data and evidence to the contrary. Neither the frequency nor the severity of these big storms is increasing.

Tropical Cyclone Terminology

A **cyclone** is an inwardly spiraling large air mass that rotates around a strong low-pressure center. Cyclones rotate counterclockwise in the northern hemisphere and clockwise in the southern hemisphere. Low-pressure centers, tropical depressions, tropical storms, hurricanes, typhoons, and tornadoes are all examples of organized cyclonic motion.

The **tropics** are a year-round hot and moist belt surrounding the equator. Its northern limit is the Tropic of Cancer at 23.437° latitude north of the equator, and its southern limit is the Tropic of Capricorn at 23.437° latitude south of the equator.

A **tropical cyclone (TC)** is a generic umbrella term for cyclones that originate in the tropics where it is sunny, and the water is relatively warm in comparison to the mid-latitude and polar bands of the planet. A TC forms as a result of Earth's motion, gravity, solar heating and evaporation of an ocean surface, and rising moist air that re-condenses with height and releases energy in the process (latent heat of condensation). This generates increasingly strong rotating winds that we then categorize as tropical cyclones according to intensity and geography. Globally, tropical cyclones originate in geographic basins. In the Northern Hemisphere, they are the North Atlantic, East Pacific, West Pacific, and North Indian Ocean basins. In the Southern Hemisphere they are the South Pacific and South Indian Ocean basins. For intensity purposes, we will focus on the North Atlantic (including the Caribbean) and East and West Pacific Ocean basins. The rest of the regions have slightly different intensity nomenclatures that are not relevant to this discussion. The intensity categories are as follows:

A **tropical depression/disturbance** is a small unnamed classification of a tropical cyclone that has a defined surface circulation and maximum sustained winds of less than or equal to (≤) 34 knots (kts), which is the same as 40 miles per hour (mph).

A **tropical storm** is a larger and uniquely named classification of a tropical cyclone that also has defined surface circulation and maximum sustained winds between 34 and 64 kts or 40–74 mph.

Hurricane and typhoons are the largest and uniquely named classifications of tropical cyclones. They have clearly defined circulations, usually with relatively calm, lowest pressure eyes at the center, and maximum sustained wind greater than or equal to (≥) 64 kts, which is the same as 74 mph. A hurricane is both a North Atlantic and Northeast Pacific tropical cyclone phenomenon. The "North" in North Atlantic refers to hurricanes that begin in the portion of the Atlantic Ocean that is north of the equator, which is where all hurricanes that affect the gulf and eastern seaboard of the U.S. begin. Northeast

Pacific means north of the equator, east of the International Date Line, and includes all of the Hawaiian Islands. A typhoon is a Northwest Pacific tropical cyclone phenomenon. Northwest Pacific means north of the equator and west of the International Date Line and includes the Philippines. Otherwise, hurricanes and typhoons have the same structural and intensity ranges.

The **frequency** of tropical cyclones is the total number of tropical cyclones that occur in any given year.

The **intensity, or severity**, of tropical cyclones is the total Accumulated Cyclone Energy (ACE) in any given year.

With the above definitions in hand, we are now ready to dive into the real, raw, no-kidding, measured and observed evidence that disproves the hypothesis that hurricanes are getting worse because of global warming.

Frequency Evidence

Figure 11-4 is a graph of the number, or frequency, of worldwide tropical cyclones observations since recordkeeping began in 1851.[4]

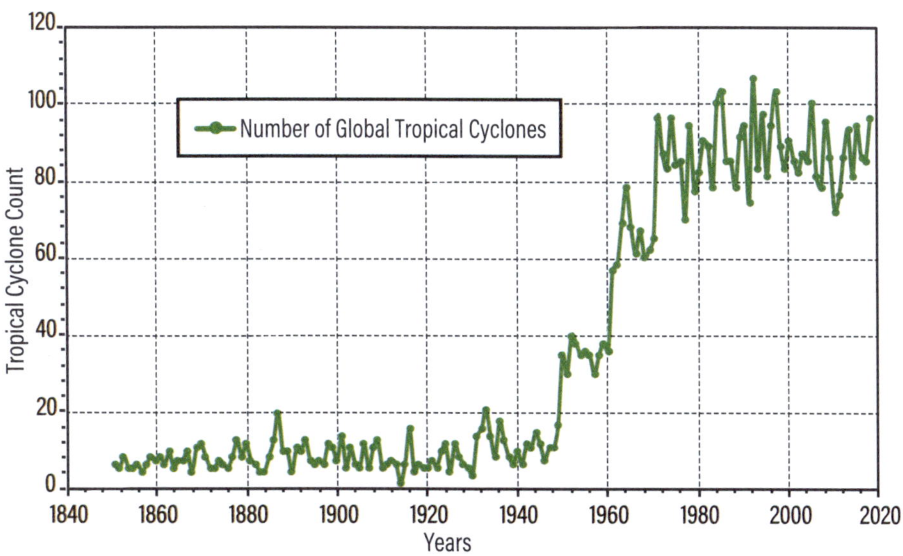

Figure 11-4. Number (frequency) of recorded global tropical cyclones since 1851.

The trend in the number of tropical cyclones reported globally was flat from 1851 to 1948 and then steadily increased until 1970. That's because all data before 1970 isn't global in nature. It is all local or regional observations, at best. Only the relatively flat trend in data from 1970 and later is truly global. This is due to improved observational technology, and specifically the steady maturation of the family of geo-stationary meteorological satellites that launched starting in the 1970s. [Geo-stationary means the satellite rotates at the same speed and in the same direction as the Earth, thus the satellite looks at the same area of Earth all the time.] By 1990, these satellites became fully operational and provided much more accurate tropical cyclone reporting on a worldwide basis. This new satellite technology for the counting of tropical cyclones on a global scale is the reason for the shift from an increasing trend to the flat frequency trend in these storms. This includes the contributions from the North Atlantic hurricanes that devastated the U.S. and Caribbean coastal regions during 2017 (Harvey, Irma, and Maria) and 2018 (Florence and Michael).

Now let's take a closer look at the relatively flat and well-monitored period between 1970 and 2018 in Figure 11–5.

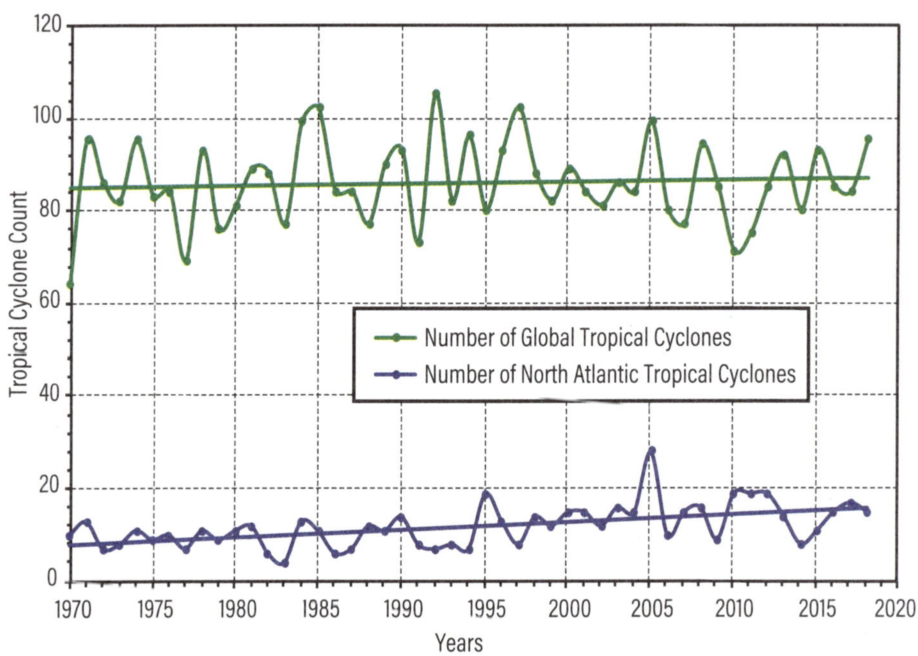

Figure 11–5. Global vs. regional tropical cyclone frequency between 1970 and 2018.

The **green** zigzag curve at the top of Figure 11–5 is the **worldwide count** of tropical cyclones. The **blue** zigzag curve at the bottom of Figure 11–5 is the **North Atlantic Basin regional count** of tropical hurricanes that affect the U.S. eastern seaboard, the Gulf of Mexico, and the Caribbean islands.

Of particular note, though, are the green and blue straight lines that are the worldwide and North Atlantic regional trends, respectively. The worldwide trend is flat (no trend), and the North Atlantic trend is slightly upward. That means that the trend in the other parts of the world (like the Indian Ocean, Australia, etc.) is slightly downward, over the same period. When you add up all of the trends, **there is no upward trend in global tropical cyclones.** And, even though there are ups and downs from time to time and region to region, the net change in global tropical cyclone frequency from 1970–2018 is zero.

These are two clear-cut, real-life examples of how **weather** (the zigzag solid curves) and **climate** (the straight trend lines) differ.

Cyclone Severity Evidence

Now we switch over to the other tropical cyclone metric, namely the intensity, or severity, of all those worldwide tropical cyclones that occur each year.[4] The way we measure severity is in terms of the instantaneous wind speed of the cyclones that is directly related to the kinetic energy within the cyclones. Modern technology advances since 1970 have been especially helpful in this regard. Specifically, the advent of pulsed Doppler RADAR instruments has enabled measurement of wind speed within tropical cyclones.

The method used to categorize the intensity of sustained hurricane level winds in the North Atlantic basin is the Saffir-Simpson Hurricane Wind Scale shown in Figure 11–6.[5]

Figure 11–7 joins two related wind scale classifications that are not part of the official Saffir-Simpson Hurricane Wind Scale, but are precursors to it.

Category	Wind Speeds (for 1-minute maximum sustained winds)			Damage
	Knots (kts)	Miles/Hr (mph)	Kilometers/Hr (kmph)	
1	64-82	74-95	119-153	Very dangerous winds and some damage to structures
2	83-95	96-110	154-177	Extreme winds and extensive damage
3	96-112	111-129	178-208	Extreme winds and devastating damage
4	113-136	130-156	209-251	Extreme winds and catastrophic damage
5	≥ 137	≥ 157	≥ 252	Extreme winds and catastrophic damage

Figure 11-6. The Saffir-Simpson hurricane wind scale.

These "related classifications" represent the two early stages of tropical storm development before growing into full-blown cyclones. All tropical cyclones begin as un-named tropical depressions. A depression is a term referring to a lower than normal atmospheric pressure region that has rotation associated with it. Once the wind speed grows to 34 kts (40 mph) the tropical depression becomes a tropical storm and is given a formal name.

RELATED CLASSIFICATIONS			
	Knots (kts)	Miles/Hr (mph)	Kilometers/Hr (kmph)
Tropical Depression	≤ 33	≤ 38	≤ 62
Tropical Storm	34-63	39-73	63-118

Figure 11-7. Precursors to the Saffir-Simpson hurricane wind scale.

Meteorological stations measure maximum sustained wind speeds of cyclones every six hours. The square of the wind speed values, summed over the life of the cyclone, is referred to as the Accumulated Cyclone Energy (ACE) metric. It is proportional to the total kinetic energy during the cyclone's lifetime.

Figure 11–8 is a graph of annual ACE for worldwide tropical cyclones between 1970 and 2018 (solid green zigzag line) and the

North Atlantic basin's contribution to worldwide ACE (solid blue zigzag line).

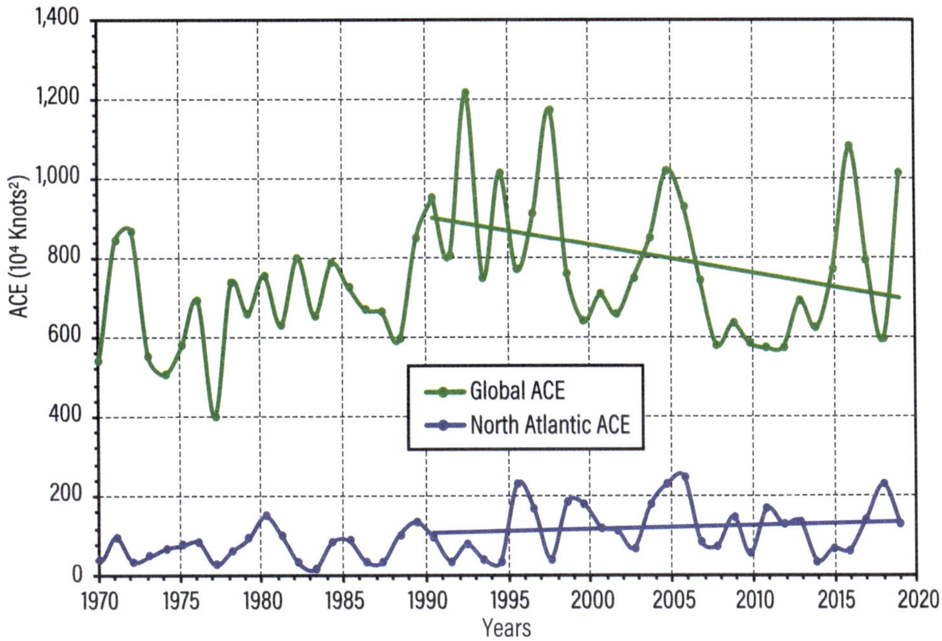

Figure 11-8. Global vs. regional accumulated cyclone energy (ACE) between 1970 and 2018.

The North Atlantic basin's ACE is the energy contribution of hurricanes that affect the continental U.S. and the Caribbean islands. The green and blue straight lines represent the ACE trend worldwide and in the North Atlantic, respectively, for the 28-year period of 1990 through 2018. ACE data previous to 1990 is not considered reliable, and therefore is not included in the trends. We see that the global ACE trend is down and the North Atlantic Basin ACE trend is nearly flat.

Today, ACE measurements are generally viewed as the most representative portrayal of tropical cyclone **severity**. However, on a global scale, we still utilize tropical cyclone frequency measurements to complement accumulated cyclone energy measurements.

The combination of frequency and ACE evidence shows that global tropical cyclones are trending down, not up.

CHAPTER TAKEAWAYS

1. Typhoons and hurricanes are unique names that fall under the umbrella term "tropical cyclones."
2. Prior to 1945, visual observation of tropical cyclones by ships at sea and observers on land were the only forms of tropical cyclone record keeping. Since then, airborne observation by Navy, Air Force, and NOAA hurricane hunters have dramatically improved position tracking and warning of these storms and hinted at their severity. But, the most significant improvement in tracking and warning arose from the deployment of geo-stationary weather satellites during the 1970s and 1980s. In parallel, the development and deployment of long-range and pulsed Doppler RADAR instrumentation enabled the measurement of cyclone severity as well as further enhanced tracking capabilities. These technology advancements have significantly improved worldwide monitoring of cyclonic weather activity.
3. What we have learned from improved global scale monitoring and data collection over the last 48 years, is that the climatic trend in tropical cyclone frequency is flat and severity has steadily decreased.
4. The evidence contradicts the assertion that tropic cyclones are getting worse. But, future evidence is necessary to further substantiate this contradiction because more (years of data) is always better. Damage and deaths also are trending downward. But, a word of caution is in order about damage and deaths. These two trends are highly variable and could easily trend upward if, for example, a swarm of tropical cyclones were to hit highly populated coastal areas. On average, they don't, but could if coastal development rises significantly in areas that are at risk of tropical cyclone activity. Where we build is the "human" factor that determines the cost to life and property.

chapter 12

More Extreme Weather-Related Events

In recent years, climate change, and specifically the global warming portion of it, has become the environmental scapegoat for any and every disaster, natural or otherwise, that befalls planet Earth. You name it—too hot, too cold, too wet, too dry, the color of the sky, the price of rice, and on and on. To begin, here are a few things to keep in mind as we delve into this subject:

We live in a solar system where we are affected by what happens around us locally as well as by the entire domain of our solar system.

Energy is conserved within our air/land/sea system. That means we neither create nor destroy energy. It is true that the energy of our air-land-water system does get mildly out of balance from time to time—more energy into our system when we are in a warming cycle and more energy out of the system when we are in a cooling cycle. But, overall long-term balance is always maintained. There is no catastrophic, runaway warming or cooling.

ACRONYMS USED IN THIS CHAPTER

Acronym	Definition
CDC	(U.S.) Center for Disease Control
CO_2	Carbon Dioxide
GWIS	Global Wildfire Information System
MSL	Mean Sea Level
MM/YR	Millimeters per Year

Acronym	Definition
MPH	Miles Per Hour
NIFC	National Interagency Fire Center
SLR	Sea Level Rise
USA49	The "Continental" United States of America

Continuous and complex fluid motion dominates our Earthly air, land, and water system. A set of nonlinear partial differential equations, known in the pillars of mathematics as the Navier-Stokes equations, governs this type of fluid motion. To give you a sense of how complex this motion can be, "Navier-Stokes existence and smoothness" happens to be one of six, still unsolved, "millennial prize problems" in mathematics.[1] Since May 2000, there has been a one million dollar prize on the table for the first person to solve this or any of the other five problems. As of this writing, the Navier-Stokes problem remains unsolved. The motion and interaction of water, land, and air, as the Earth repetitively spins, orbits, tilts, and precesses, are definitive examples of the Navier-Stokes problem. This is the scene of a continuing battle for better understanding.

Water drives the "greenhouse effect" that acts as a blanket to ensure our comfortably warm, rather than freezing, climate.

Water and carbon dioxide also drive the photosynthesis process that produces the food we eat and the majority of the oxygen in the air we breathe.

Weather events are not the same as climate. A series of short-term weather phenomena over time and space add up to define a cyclic, energy-conserving global climate.

Storms are not becoming more frequent or severe.

Chapters 10 about tornadoes and 11 about hurricanes covered the variable, complex, and recurring nature of these two weather phenomena. I'll now incorporate other prominent, environmental phenomena that are **supposed** to be happening with **alarming frequency** these days. Well, at least they are reported with alarming frequency by the media. I'm talking about wildfires, droughts, heat waves, and rising sea levels. The common, knee-jerk reaction to such phenomena is familiar. Activists insist that these events are a result of global warming, caused by greenhouse gases, with CO_2 as the main culprit, and that these events are the by-product of humankind's insatiable appetite for fossil fuels. Furthermore, their conclusion is that the whole world is going to burst into flames at any minute as a result. Not true. Empirical evidence shows that reaction to be incorrect.

Wildfires

There are two major sources of wildfire data. The worldwide database is the Global Wildfire Information System (GWIS), whose wildfire data spans the period 2001–2017.[2] Covering the continental United States (USA49) is the National Interagency Fire Center (NIFC), whose reliable data spans the period 1985–2019.[3] Both databases show similar trends in wildfires. Given the angst about devastating wildfires in the western U.S., we'll confine our analysis to the 49 U.S. continental states (USA49) database.

Lightning, arsonists, campers, hikers, hunters, winds, downed power lines, improper environmental policy, and on and on, all cause wildfires or contribute to their severity. But warming temperature is not a major contributor.

In fact, while global temperatures are up slightly—just under 1 °F in the last 34 years—wildfires are down slightly. The same is true for the USA49. However, on a regional level within USA49, there are localized places, like California, Oregon, and Washington, where both dryness and wildfire frequency commonly increase in the fall. However, the frequency of such events is not increasing overall.

Of profound concern in November of 2018 were the horrific Camp and Woolsey wildfires of Northern and Southern California, respectively. Global warming was not a significant factor in the Camp Fire[4] or any of the other frequently occurring Northern or Southern California fires for that matter. Warming had little to do with the dry conditions in the region. In the aftermath of the Camp Fire, it has become clear that the severity of such fires is a combination of several factors. They include poor urban planning, needlessly restrictive environmental policy over our vast forests, the high cost of preemptively trimming underbrush, the fundamental flammability of the grasslands, and early on inaccessibility to the fires.

Most importantly though, there also is a natural phenomenon that greatly augments the severity and burnt acreage of virtually all wildfires in California. This natural phenomenon is known as "offshore downslope wind."[4] These are the infamous Santa Ana winds in Southern California and Diablo winds in Northern California. In the fall season, a natural temperature difference (aka gradient) between the cold desert area of Nevada and the warm coastline of California generates these east-to-west (easterly) winds coming over the mountains. The winds then speed up further as they flow downslope toward the Pacific Ocean. These Santa Ana and Diablo winds played a huge part in the Woolsey fire in Southern California and the Camp Fire in Northern California.

Ironically, a 2009 climate **model-based** study[5] determined that global warming should **DECREASE** the severity of Santa Ana (and presumably Diablo) winds, certainly not what we read or hear. Who knew that warming helps dampen wildfires!

Switching now to a national perspective, Figure 12–1 illustrates the profile and trend of continental U.S. temperatures, as measured by satellite during the period of 1985–2019.[6] The zigzag blue line represents lower tropospheric (near surface) annual temperatures and the straight black line is the linear trend of temperatures over the entire time period.

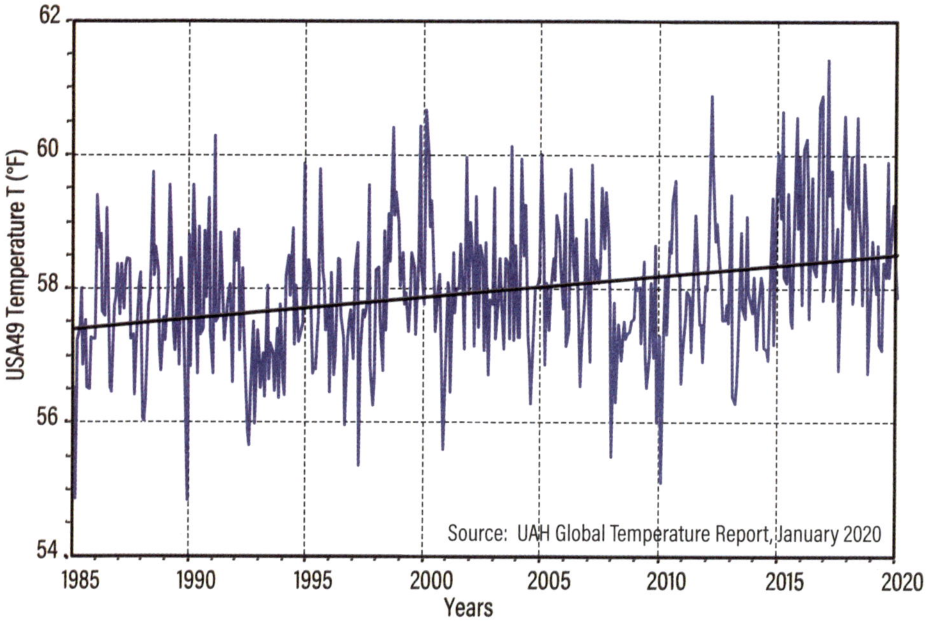

Figure 12-1. Continental United States (USA49) temperature profile and trend.

Figure 12–2, below, is the corresponding annual profile of the number (frequency) of wildfires, in red, and the trend in black, that occurred in the continental U.S. during 1985–2018 time frame.[7]

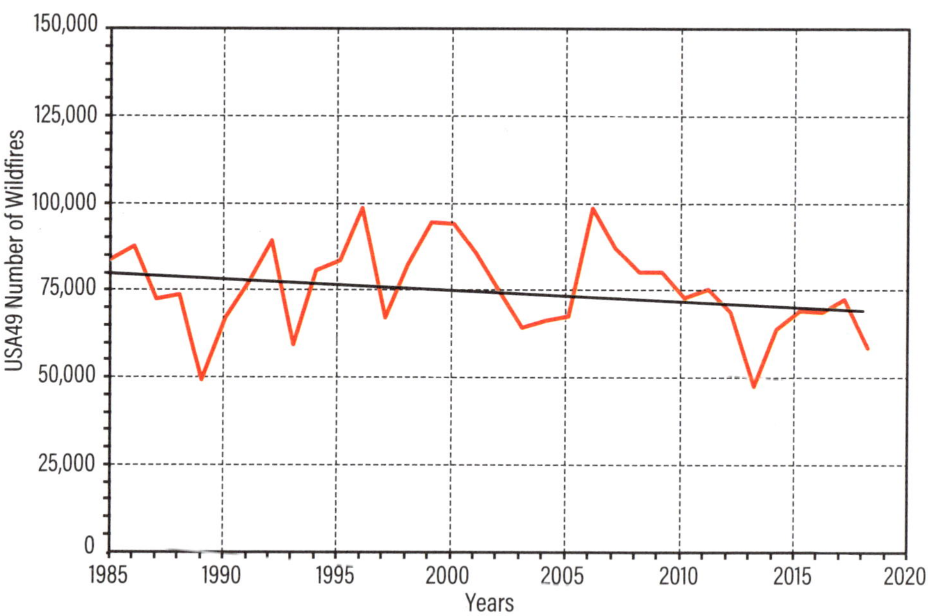

Figure 12-2. USA49 wildfire frequency profile.

The messages from figures 12–1 and 12–2 are that U.S. temperature has trended upward by slightly less than 1 °F in 34 years, and the number of U.S. wildfires has trended downward by about 25,000 in 33 years. This clarifies that near surface temperatures are not a driving force for the number of wildfires in the U.S.

You might question why only the period of 1985 through 2018 is shown. Some would call that blatant "data cherry-picking," since USA49 wildfire records actually go all the way back to 1926, as shown in Figure 12–3.

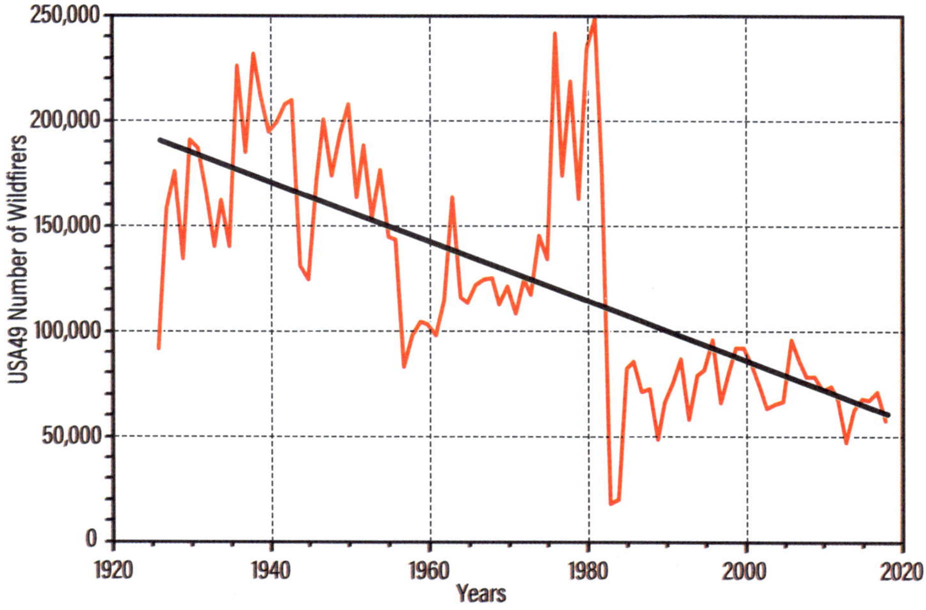

Figure 12-3. Expanded USA49 wildfire frequency profile.

This expanded (1920–2018) version of U.S. wildfire history shows a much steeper decline in wildfires than in Figure 12–2. However, the National Interagency Fire Center (NIFC) regards data prior to 1983 as unreliable because "sources of these figures are not known, or cannot be confirmed, and were not derived from the current situation reporting process."

As a result, and even though normally I'm a big picture guy, I chose to analyze only the shorter 1985–2018 range of frequency data in Figure 12–2 and disregarded the pre 1985 data in Figure 12–3.

Given the circumstances, that was the right thing to do, and is not cherry picking of short time-frame data. The best data available was used.

Another wildfire metric is **acres burned.**[7] The trend in the number of wildfires has not increased in the U.S. during the reliable reporting period of 1985 to the present. However, the trend in the number of acres burned by wildfires has trended substantially upward, from roughly 2.5 million to 8.5 million acres burned since 1985. This is a useful metric because it accounts for important variables that include fuel content available to the fire, local weather conditions, and resources available to fight the fire.

To conclude this wildfire section, there is one more interesting metric (Figure 12–4).

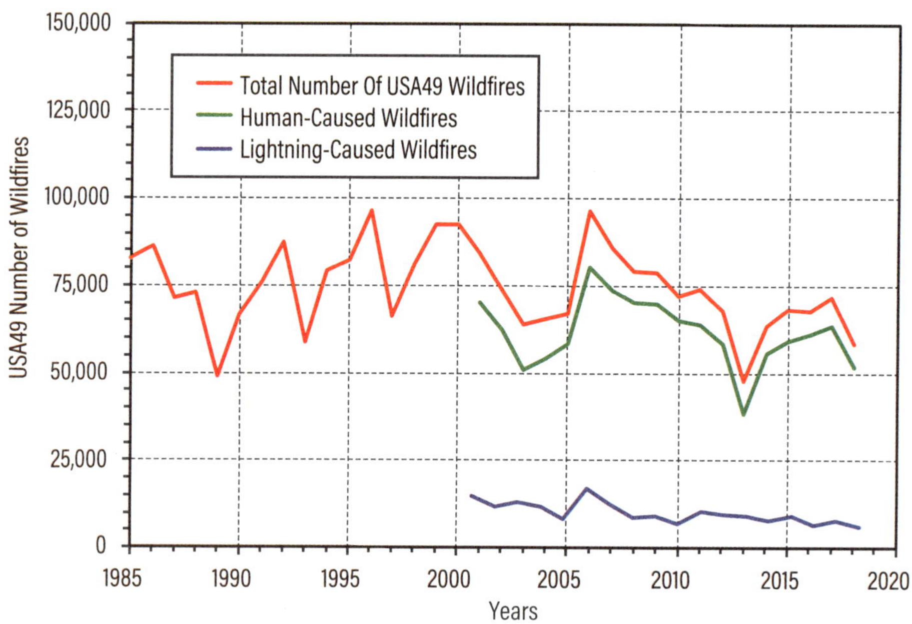

Figure 12–4. Major contributors to USA49 wildfires.

It is the total number of USA49 wildfires again. But this time, I've added two subcategories to the total wildfire graph, namely human-caused wildfires (the solid green line) and lightning-caused wildfires (the solid blue line).[8]

The green and blue lines add up to the red line for the years 2001 through 2018. The green and blue lines don't start until 2001 because the NIFC did not begin to distinguish wildfires by cause until 2001. Since then, human-caused wildfires are the vast majority of the total, with lightning-caused fires being less than twenty percent of the total. These human-caused numbers include intentional (arson) fires, as well as unintentional fires caused by campers, hikers, power lines, and the like. They do not include fires caused by carbon dioxide in the air. And that is as it should be, because CO_2 is not flammable. Most importantly, fossil fuel generated CO_2 does not absorb nearly enough heat via the greenhouse effect to sway wildfire numbers in the least. It would be deceptive, at best, and an out-and-out lie, at worst, to blame wildfires on CO_2. Yet, this quite commonly mistaken idea still refuses to be extinguished.[9]

Heat Waves and Droughts

Figure 12–5 illustrates historical heat wave index for the continental United State of America (USA49) since 1900.[10]

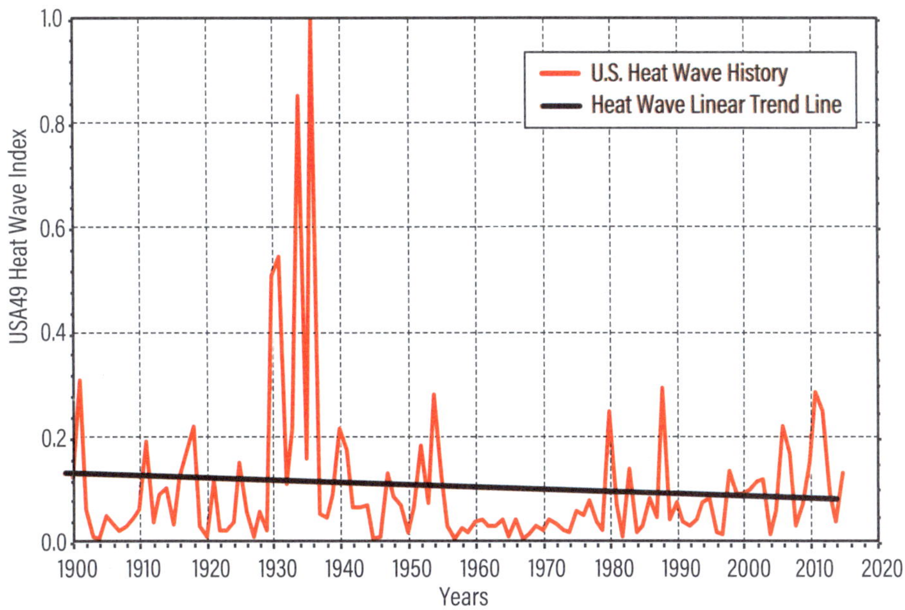

Figure 12–5. USA49 heat wave index and trend line.

The zigzag red line and black trend line represent heat waves in the U.S. since 1900. There were a lot more heat waves during the 1931–1936 dust bowl era than any other recent time, before or after. The numbers of heat waves, as well as their durations, have trended steadily downward.

Note that the heat wave "index" as plotted in Figure 12–5, is the product of the land area percentage covered by the heat wave and the frequency of occurrence of the heat wave. For example, a heat wave index value of 0.4 in any given year could represent 40% of the U.S. experiencing one heat wave, or 20% of the country experiencing two heat waves, or any other combination.

Drought conditions, from abnormally dry to extreme and exceptional drought, have been relatively flat to slightly downward since 1900,[11] as shown in Figure 12–6.

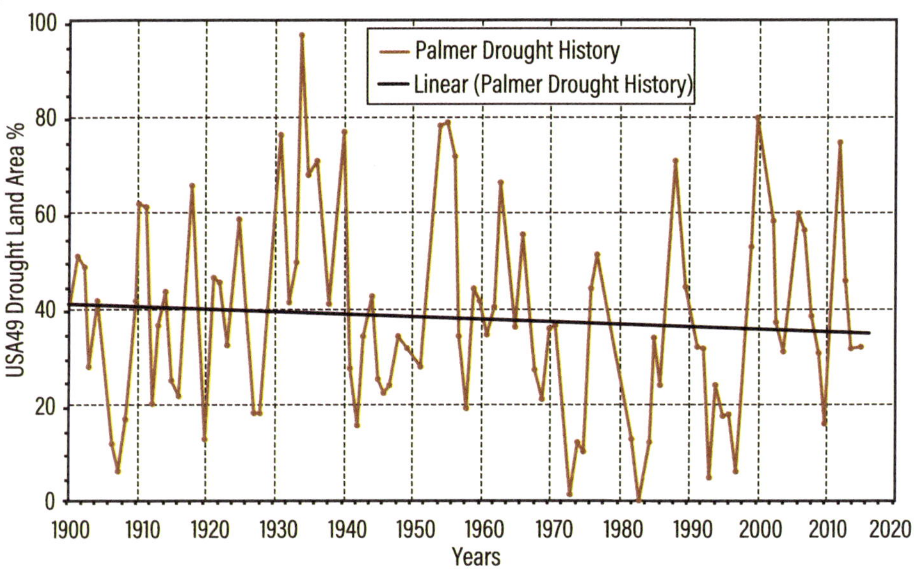

Figure 12-6. USA49 Drought land area percentage and trend line.

The brown zigzag line in Figure 12–6 represents the annual drought trend in terms of percentage of land area affected by droughts since 1900.

The number of all source deaths attributable to heat is also trending slightly downward. In contrast, the number of all source deaths attributable to cold is relatively flat during the same period. But the more interesting statistic from the U.S. Centers for Disease Control (CDC)[12] is that there are about 17.6 times more deaths annually in the United States from hypothermia (too much cold), as there are from hyperthermia (too much heat) since 1995. Worldwide heat waves, drought, and related death metrics are consistent with United States metrics. Perhaps we should be worrying more about cold waves than heat waves.

Sea Level Rise (SLR)

What about rising sea levels and flooding? First, let's separate the terms sea level rise and flooding. Just as I have harped that weather and climate are not interchangeable terms, neither are Sea Level Rise (SLR) and floods. SLR is associated with a **long-term** increase of ocean levels, relative to land levels, that are caused mainly by melting ice, volcanic activity and Earth's tectonic motion. In contrast, floods are associated with **shorter-term** overflow of water that submerges usually dry land. An example would be the Mississippi river flooding in Iowa in the spring.

Sea levels are currently rising at a slight but steady rate of about **two millimeters** per year (approximately 1/16 inch).[13, 14] The fundamental reason for SLR is that we are in an interglacial warming period. Three forms of SLR happen as a result:

1. The interglacial warming expands the volume of ocean water, which raises the sea level. We call this "thermal expansion" of ocean water.
2. Interglacial warming melts land-based glaciers over time. Some of this continental water runoff flows into the ocean to somewhat raise sea level.
3. Floating ice in Arctic and Antarctic waters melts over time and raises the sea level a small amount. How much sea level increases

has to do with the difference in salinity between sea ice and surrounding seawater.[15] If the ice and water salinity are the same, then there is no sea level rise. That is the familiar case we know happens when an ice cube melts in a glass of tap water—no change in the water level as the ice melts.

So, what would the impact be if all Arctic sea ice were to melt? The current estimate of all floating sea ice is about 660,000 cubic kilometers (158,342 cubic miles). If all of it were to melt completely, then the global sea level would raise about 1.5 inches (40 millimeters) above its current level. That's a **worst** case. And 1.5 inches would **NOT** be catastrophic. Anybody that can't adapt to a 1.5-inch rise in sea level deserves to get wet.

Figure 12–7 shows total sea level rise (SLR) in **millimeters** since the end of the last major ice age[16] approximately 21,500 years ago until today (the zero mark at the right hand edge of the horizontal axis).

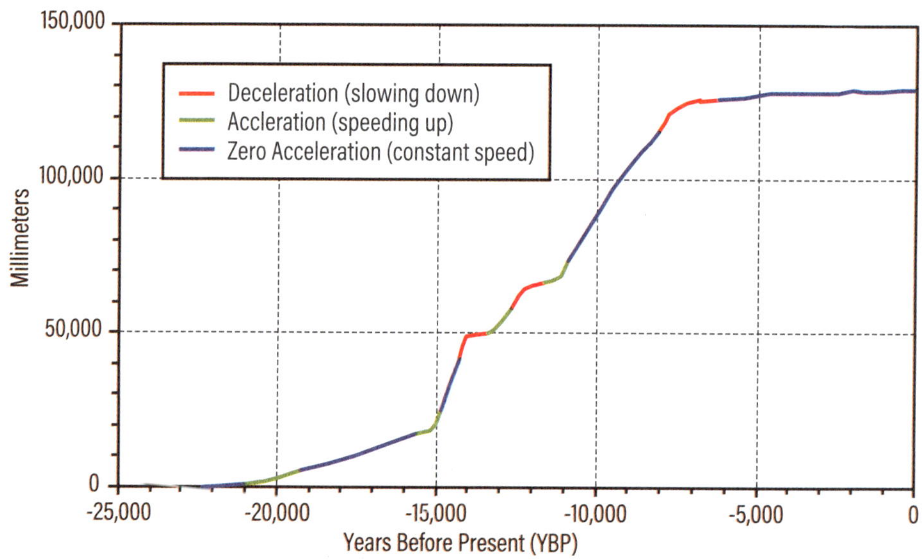

Figure 12-7. Sea level rise since the last major ice age.

Periods of accelerating SLR are as indicated by the green concave curved segments of Figure 12–7. Periods of decelerating SLR are the red convex curved segments. And the rest of the relatively flat (non-curving) blue line segments of the graph are times of constant rise of sea level (zero acceleration and zero deceleration). A good analogy would be driving your car. Let's say you are at a stop on a country road. You step on the gas to go from 0 to 60 miles per hour (mph). When you do this, you are accelerating. Then for a while you travel along at a steady speed of 60 mph—no speeding up or down. That is zero acceleration and zero deceleration. Next, you come upon a small town and apply the brake to slow down to 30 mph. That process of going from 60 down to 30 mph is deceleration.

The most recent segment of sea level deceleration, shown in Figure 12–7, occurred around 7,500 to 6,500 years ago. Sea level rise for the last 6,500 years has been at a relatively steady (i.e., zero acceleration) rate of approximately 2 mm/year, which is just over 1/16th of an inch per year. At that rate, coastal areas of California, or New York, or Florida, or Pacific islands won't be underwater anytime soon. It would take almost 19,000 more years for you to be walking around Orlando, Florida's Disneyworld knee-deep in seawater. And that would be in the highly unlikely case of no tectonic motion. Tectonic motion causes irregular and unpredictable sea level rise or fall. So, the truth is we scientists really don't know what will happen with regard to sea level rise in the future.

However, if you believe, as I do, that history tends to repeat itself, then ice will likely continue to melt and sea levels will continue to rise at a zero-acceleration rate of about two mm/year, until the end of our current interglacial warm period. Then, Earth will drop down into another major ice age—which means Disneyworld is safe for the time being. The whole process of glacial cooling followed by interglacial warming will repeat continuously. The only thing we can do about it is to prepare for and adapt to both warming and cooling.

Flooding

Floods are an overflow of water that submerges usually dry land. Hurricanes near coastlines, severe rainfall, substantial snowmelt, interglacial sea level rise, sporadic tidal waves (aka tsunamis), and other wave motion are all sources that contribute to flooding.

Severe rainfall flooding typically originates from short-term, severe weather phenomena. An example of this was the devastating hurricane "Sandy" that hit the New York City area in October 2012. Hurricanes that rake the North Atlantic basin have fluctuated up and down over the past 49 years, with a slightly increasing overall trend in frequency and severity over the period, but not by much. And from a **global** perspective, the number of cyclones is flat (unchanged) and severity (aka ACE) is decreasing. See Figures 11–5 and 11–8 in Chapter 11.]

Reasons that sea level rise may cause flooding include:

4. Global circulation of ocean water, i.e., ocean currents due to the Earth's spin;
5. Sinking and expanding of large land masses, i.e., plate tectonics; and
6. Melting of ice from the Arctic. This is a feature of long-term and periodic interglacial warm periods like the one in which we now live.

The above characteristics of sea level rise are demonstrably natural consequences of the sun, orbital motion, inertia, and, most importantly, the amount of water in our air/land/sea system. All are clearly beyond human control, as illustrated and substantiated by empirical evidence presented in the previous SLR section.

Environmental and climate change reports blame human-caused, carbon dioxide driven warming for the flooding effects of hurricanes like Florence in 2018.[17] But historical evidence says otherwise. Temperature drives carbon dioxide (see Chapters 8 and 9), not the other way around. And, just as important, we are in the latter stages of an interglacial warming period (see Chapter 6, Figure 6–6) where melting is a fact of life that won't change substantially until planetary motion brings on the next ice age.

In summary, there is no long-term global trend in flooding. It's been flat over many years. It's like a game of whack-a-mole—flooding here, drought there, and on and on.

So, why is it that supposedly 500-year or 100-year flood events are all of a sudden hitting places like New York City more often? Why is the "mole" picking on them?

Perhaps it has to do with a misunderstanding of the terms such as 500-year or 100-year flood events. For example, the U.S. Geological Survey describes 100-year flood event as a "flood having a 100-year recurrence interval . . . In other words, a flood of that magnitude has a one percent (1/100) chance of happening in any year."[18] This does not mean that when it does happen, it shouldn't happen again for another 100 years.

A December 2018 Colorado State University report[19] gives the probability of a North Atlantic hurricane impacting New York as eight percent per year. This probability is based on the U.S. National Hurricane Center's historical impact database. That's a lot more than a one percent chance per year. To me, the Hurricane Center's report suggests that New York City's prediction of their flood recurrence interval might need to be updated.

Additionally, there are a number of recent media reports that cite studies by scientists at Penn State University[20], Massachusetts Institute of Technology[21], and others[22] that conclude floods will be hitting New York City more often in the future. Based on historical evidence, I believe they are mistaken so far. If history continues to repeat itself, they are mistaken about the future as well. But that's okay. Science is loaded with trial and error.

Whatever the hypothesis, I do not believe we have enough information or knowledge to create computer models that predict run-away global warming. And, we humans are not in any way able to control the tremendous natural forces that rule our climate. Reality rules the day, not computer climate simulations.

But, what will happen in the future is a legitimate question. We do not know the answer. Why we don't know what will happen in the future, with regard to flooding, is not so simple. It's about complicated fluid dynamics, gravity, inertia, etc.

Nonetheless, sensational horror stories that say New York City, Miami, and other coastal cities will be underwater continue to be favorite themes of the mainstream media and movies.[23] Sensation sells, regardless of reality.

CHAPTER TAKEAWAYS

1. Global and regional (U.S.) surface temperatures are up about 1 °F over the last 35 years and carbon dioxide levels are up about 65 ppm over the same period. However:
   a. Continental USA wildfires are down by about 25,000 in number since 1985, but acres burned are up by roughly six million acres since 1985.
   b. Heat waves and droughts are flat to slightly down since 1900.
   c. Sea levels are slightly rising at a steady rate of 2–3 millimeters per year. They are natural phenomena that likely will continue until the Earth's tilt and eccentricity realign and we descend into another ice age. Until then, we must be prepared and able to adapt to slight sea level rise.
   d. Large-scale coastal rainfall and flooding coincide with the number of cyclones. Both are unchanged over time. Coastal rainfall and flooding do not coincide with cyclone severity, which is decreasing over time. Non-coastal river flooding tracks seasonal snowmelt and irregular and heavy rainstorm activity.

2. None of these things happen by accident. And they don't happen because of climbing temperatures or CO_2 levels. They do happen because of an energy-balancing act that occurs constantly throughout our air/land/sea system. The end result is the presence of a "Goldilocks-like" not too hot and not too cold interglacial warm period, caused by the interaction of multiple, monumental forces, namely tectonic motion, orbital motion, inertia, the sun, and water. All are way beyond human influence or control. Any self-proclaimed pundits, who suggest otherwise, typically with scare tactics, are wrong.

chapter 13

Summary

What Do We Do Now?

Our atmospheric system is working as intended—in a balanced and cyclic way. Water and carbon dioxide help nourish and sustain all life on the planet and provide a comfortable, warm greenhouse blanket. As our miraculous Earth orbits our sun and moves in the expanse of space, the climate warms and cools and we adapt to that variability.

What can we do to help ensure that life on our planet continues to adapt and thrive? Here are some ideas, beginning with general ones, and then getting more specific:

Minimize air, land, and water pollution. Good stewardship of our planet is paramount, and everyone's continuous responsibility. Do the most beneficial things, based on known science and data, not politics or scare tactics.

Control pollution, not CO_2. Be careful to not equate pollutants and greenhouse gases. The primary greenhouse gas is water vapor. Carbon dioxide and methane are in a far distant second and third place. Water vapor and CO_2 are colorless, odorless, non-toxic greenhouse gases that are vital for all life. They are not pollutants. The main pollutants in our air are ozone, nitrogen and sulfur oxides, ammonium nitrate, carbon monoxide, volatile organic compounds, smoke, dust, and black carbon. Over the years we have made substantial strides to reduce these real pollution culprits. But some of these pollutants still linger as visible, toxic reminders that we still need to do more, especially in our cities and suburbs. This is where we need to focus our blood, sweat, tears, and limited funds.

Embrace CO_2. Forget ridding ourselves of CO_2 via carbon tax or sequestration to prevent supposed catastrophic warming. Increasing CO_2 is not a problem—it's beneficial.

Innovate, plan and adapt. Our climate is variable and always has been. We need to recognize that nature, not humans, controls our earthly climate. We need to be prepared for its ups and downs and not fantasize about changing it to our liking and succumbing to baseless fear mongering. Taxation and uninformed regulation are not good ways to prepare for heat or cold.

Beware of "crystal ball" computer models that forecast and predict the future. To have any validity, a computer model must be independently reproduced (verified) and independently matched (validated) to empirical data (evidence). Scientific models can be helpful, but only if the inputs and assumptions to, and the physics within, the models are accurate, and not manipulated to generate a desired result. Climate models must correctly represent real atmospheric processes.

Convince our government leaders to conduct robust, truth-based, public debates that address all sides of climate science.

Insist that policy decisions be based on cost-benefit, life-cycle analyses. This will help in making wise decisions about what to do with our nation's money.

Support research, development, and use of affordable, efficient, and aesthetic forms of energy. That means honestly explore and research all fossil, nuclear, wind, solar, hydroelectric, geothermal, biofuel, and yet undiscovered energy sources.

Continue to increase the combustion efficiency of fossil fuels being used in vehicles, power plants, and industry. Doing so will further reduce the amount of incompletely burnt residues of the combustion process. Examples of reducible residues include black carbon exhaust from diesel engines, leftover carbon monoxide that is not converted to carbon dioxide, and other volatile organic compounds from the combustion process.

Develop better batteries for storage of energy from intermittent wind and solar sources. Battery technology and miniaturization has come a long way in the age of computers, cellphones, and other electronic gadgets. Some progress has been made in car battery technology as well. But not nearly enough improvement has been made to date with regard to batteries for large-scale solar and wind applications. Energy storage is a continuing weak link preventing wind and solar power from becoming reliable and cost effective energy sources. Remember, intermittent wind and solar power are not yet, and may never be, the best answer.

Develop better energy integration and distribution processes so that electricity can move rapidly and cost effectively over large distances in order to address local lack of wind and sunlight issues. Are power towers and lines through forests, or across deserts the best we can come up with?

Build more dams and reservoirs, especially in drought and fire prone areas. Find ways to trap seasonal floodwater for drought, fire mitigation, and hydropower, rather than simply waste the water by allowing it to flow unchecked into the oceans. For example, the last dam built in California with a reservoir capacity exceeding one million acre feet was in 1979.

Relax regulations that stymie dam and reservoir construction, forest thinning, and removal of brush that fuel wildfires.

[Note: On Friday, March 22, 2019, California Governor Gavin Newsom declared a "state of emergency" with regard to wildfires and unnecessary fire suppression in California. By relaxing and waiving some environmental regulation, the Governor paved a way to improve forest and grassland health statewide. This will allow experts to initiate actions to clear dead trees, reduce the density of trees and undergrowth, and conduct prescribed burns to reduce wildfire risk. This is a wise step since trees usually leave the forest in one of two ways—as lumber or smoke.]

Build more roads and firebreaks to improve accessibility to fire prone regions.

Reduce methane leakage from natural gas extraction processes.

Enact laws and spend our national treasure based on real, proven, and validated science.

Do not get tricked by powerful groupthink. Seek truth through evidence and experience.

Do not be bullied into silence. Skepticism is the lifeblood of science.

Last, but extremely important: Keep learning. Teach our children the scientific method, logic, and history. These skills are fundamental to accurate information gathering, critical thinking, and understanding and truth.

Final Thought

For months, I've been thinking about how to end this book with a bang—a statement that summarizes the heart of my book "to a tee." And I think I've got just the thing, thanks to a clever graphical format used by a colleague, Steve Goreham, in a recent lecture of his.[1]

As you know by now, I've harped incessantly that weather and climate are not the same. Figure 13-1 is a clear example of that stark contrast.

The **multicolored** data points are **local** surface temperature readings from a single weather station—the "Snake Creek Powerhouse" weather station[2] near my home in Midway, Utah. Each monthly vertical column represents 100 years worth of data for a given month. For example, the green dots in the July column near the top center of the plot are the annual average temperature readings in Midway for each year from 1919 to 2019. Note the black arrow that points to the temperature value near the right hand edge of the July column. That

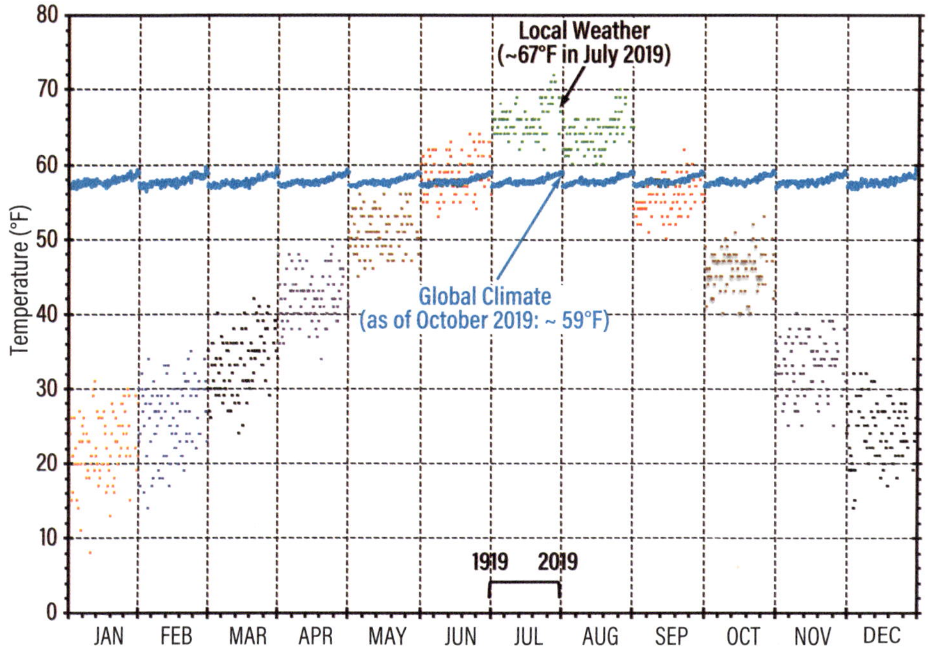

Figure 13-1. Local and global average monthly temperature variability (1919–2019).

black dot represents the average temperature at the local Midway station in July 2019. The value is approximately 67 °F. While 67 °F may seem low for a summertime July measurement, it really isn't. That's because these monthly temperatures are averages of both daytime and nighttime measurements.

The other green dots within the July column represent the earlier 99 years worth of July average temperature measurements. Each of those green dots is a Midway **weather** event.

The **blue** data are simply the average of all of the worldwide local weather stations' temperature data,[3] including Midway's. The squiggly blue lines represent monthly global climate during the last 100 years, which currently is near 59 °F.

We have experienced only about a 2 °F warming of our global climate from 57 °F to 59 °F over the past 100 years. Most people cannot even feel that small of a temperature difference. But we sure do feel the temperature differences in local, seasonal weather. Weather and climate can be quite different, both geographically and over time, when you add up all the locales over many decades.

The point here is that daily, monthly, or even yearly local temperatures vary quite widely, as illustrated by the multicolored weather data in Figure 13–1. But the variance is cyclic in nature. And when you add up all the locales over many decades of time you get nearly flat, stable, normal, steady-as-a-rock temperature climatic behavior, like the blue data in Figure 13–1. This striking result is not magic, an accident, or a coincidence. Our solar system is complex, self-adjusting, and controlled.

My concluding **opinion** is that it would be an outrageous waste of money to have a national (or international) policy that spends **multiple trillions of dollars per decade** in an attempt to reverse a global climate pattern that is natural, normal, stable, and controlled by the sun, water, gravity, and inertia. Furthermore, mankind's personal, vehicle, home, and industrial carbon dioxide emissions are miniscule, and neither concerning nor catastrophic. In fact, quite the opposite is true. One more time: carbon dioxide is a **colorless, odorless, clean** atmospheric gas that is increasing in a steady and beneficial manner, and is crucial to all life on Earth.

Yes, we must be good stewards of our air, land, and seas. Therefore, our government should spend **reasonable** amounts of money on things that we are able to control. That would be to clean up, or at least minimize, things like soot, smoke, dust, and other toxic pollutants in the air, as well as trash in the water and on the land.

Our climate is not anthropogenic (human-caused), and it's not catastrophic (burning out of control). It is quite stable and mild. And that's the heart of my book, "to a tee."

Scientific knowledge can help our government spend money on real, rather than imaginary, problems.

NOTEWORTHY BOOK TAKEAWAYS

1. CO_2 is a clean, non-toxic gas, just like water vapor. Neither are pollutants. Both are greenhouse gases.
2. Water vapor is, by far, the dominant greenhouse gas, not CO_2.
3. The sun, water, gravity, and inertia are the control knobs of our climate, not humans.
4. Based on eons of historical evidence, today's global climate is quite normal, quite stable, and quite hospitable to all forms of life.
5. Science is never settled. Any scientific statement is only one reliable, reproducible experiment away from being refuted. Ask Albert Einstein.
6. Climate science is complex. There are aspects of it we do not understand well at all yet. That severely limits the reliability and utility of any computer model or prediction of future climate.
7. There is no place for consensus in real climate science or any science for that matter.
8. The Earth is NOT warming in a catastrophic, human-caused manner. The greenhouse effect, as currently understood, helps keep Earth from freezing.
9. We all must be good stewards of our planet.
10. We must teach and challenge our kids to think critically and study the many facets of scientific information, and not just follow the mob of popular opinion.

appendix A

Apple-Earth Comparison

If you could shrink the Earth down to fit in the palm of your hand, its atmosphere would be as thin as the skin of an apple.

Figure A-1. Atmospheric shell and apple skin comparison.

The Earth's diameter (d) is approximately 7,926 miles,[1] and the thickness (h) of the light blue atmospheric shell that surrounds

Earth is approximately 52 miles thick out to the Kármán Line, which is the beginning of outer space.[2]

For comparison, the ratio of the atmospheric shell's thickness to the Earth's diameter is about the same as the ratio of an apple's skin to the diameter of the apple.

$$\frac{h_{atmosphere}}{d_{Earth}} = \frac{52 \text{ mi}}{7{,}926 \text{ mi}} = 0.007 \qquad \frac{h_{apple\ skin}}{d_{apple}} = \frac{0.5 \text{ mm}}{70 \text{ mm}} = 0.007$$

appendix B

Sun Power

On average, the sun's luminosity (Ls) is approximately joules of energy per second, or Watts (W) of solar power.[1] But how much of that solar power is intercepted by planet Earth? The geometry of Figure B–1 will help us answer this question.

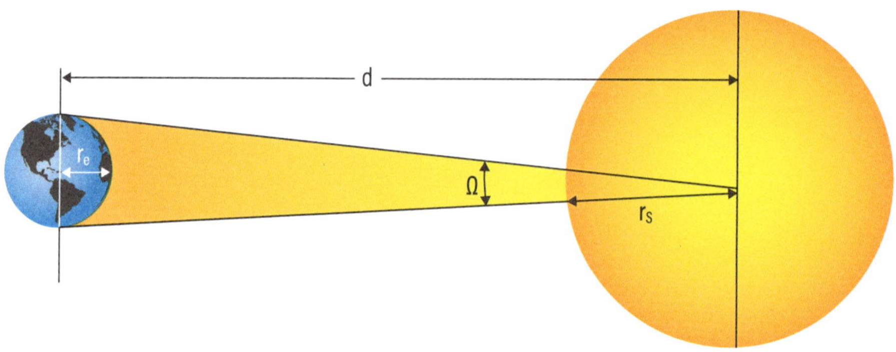

Figure B-1. Earth-sun geometrical dimensions (not to scale).

- "r_e" is the spherical radius of the Earth[2] in meters (m): $\approx 6.378 \times 10^6$ m
- "r_s" is the spherical radius of the sun[2] in meters: $\approx 6.957 \times 10^8$ m

ACRONYMS AND SYMBOLS USED IN THIS APPENDIX

Acronym	Definition
Ae	(cross-sectional) Area of the Earth
d	distance between centers of the Earth and sun
Eo	Total solar irradiance.
$\bar{E}_e$	Average irradiance in watts per square meter (W/m^2)
Is	Solar radiation intensity in Watts per steradian (W/sr)

Acronym	Definition
Ls	Luminosity (power) of the sun in joules per second
Ω	Omega (a value of a solid angle)
r_e	radius of the Earth in meters
r_s	radius of the sun in meters
sr	steradian (a unit of solid angle)
W	Watts (or Joules per second) of solar power

- "d" is the center-to-center distance from Earth-to-sun[3] in meters: $\approx 1.496 \times 10^{11}$ m

The sun emits its power "spherically and equally in all directions," which is the equivalent of 4π steradians (sr) of total solid angle. So, the total solar power emitted into any given solid angle, aka the radiant intensity of the sun (Is), is:

$$I_S = \frac{L_S}{4\pi} \cong \frac{3.828 \times 10^{26}\ W}{4\pi\ sr} \cong 3.046 \times 10^{25} \cong \frac{W}{sr}$$

From the Figure B–1 diagram, omega (Ω) is the conical solid angle that bounds the amount of solar power that the Earth receives. The cross-sectional area of the Earth (Ae) and the inverse square of the distance (d) between the Earth and the sun determine the value of omega (Ω) as follows:

$$\Omega \cong \frac{A_e}{d^2} \cong \frac{\pi r_e^2}{d^2} \cong \frac{\pi\,(6.378 \times 10^8)^2}{(1.496 \times 10^{11})^2} \cong 5.7 \times 10^{-9}\ sr$$

So, the maximum power density, or total solar irradiance (TSI), that intercepts the top surface of our Earth's atmosphere that is perpendicular to the incoming solar energy[4] is:

$$S_0 \cong \left(\frac{L_S}{4\pi}\right)\left(\frac{\Omega}{A_e}\right) \cong \left(\frac{L_S}{4\pi}\right)\left(\frac{A_e}{A_e \times d^2}\right) \cong \frac{L_S}{4\pi d^2} \cong \frac{3.828 \times 10^{26}}{4\pi\,(1.496 \times 10^{11})^2} \cong 1{,}361\ \frac{W}{m^2}$$

However, the arriving sunlight bathes the entire hemisphere of the Earth that faces the sun. Therefore, the average solar power density that bathes the entire top surface of the Earth hemisphere facing the sun is approximately one fourth of S_0[4], as follows:

$$\bar{E}_e \cong \frac{1.361}{4} \cong 340\ \frac{W}{m^2}$$

Appendix C

Gravity

Newton's "Universal Law of Gravitation,"[1] states that the larger sun attracts the smaller Earth toward it, as in Figure C–1.

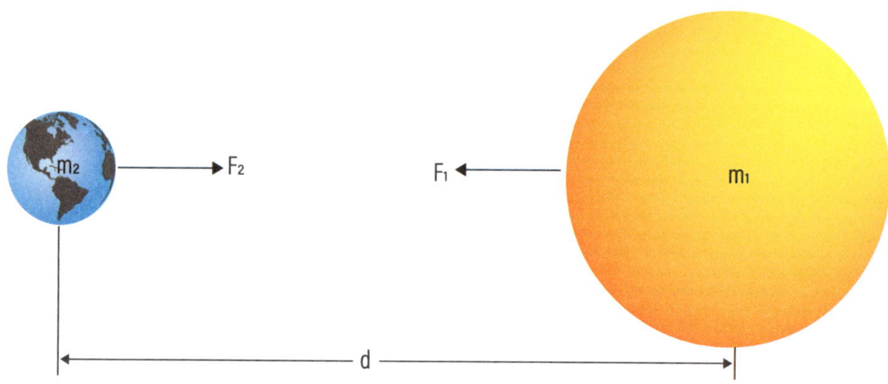

Figure C-1. Two-body gravitational attraction.

$$F_1 = F_2 = G\left[\frac{m_1 \times m_2}{d^2}\right]$$

F_1 = gravitational force attracting the sun toward the Earth;

F_2 = gravitational force attracting the Earth toward the sun;

G = universal gravitational constant (6.67×10^{-11} N $\times$ m^2/kg^2);

m_1 = mass of the sun;

m_2 = mass of the Earth;

d = distance between the Earth and the Sun.

The above relationship holds true regardless of the magnitude of the masses (m_1 and m_2) or the distance (d) between the objects. This "two-body" relationship is a subset of the much more complex "many-body" solar system interactions.

appendix D

Earth's Orbital Eccentricity

Figure D–1 shows the tilted and spinning Earth rotating around the sun. The shape of the Earth's orbit varies in time from a slightly elliptical orbit (the solid black orbital line) to a more elliptical orbit (the dashed black orbital line).[1] We call this deviation from the established circular rotation pattern "eccentricity."

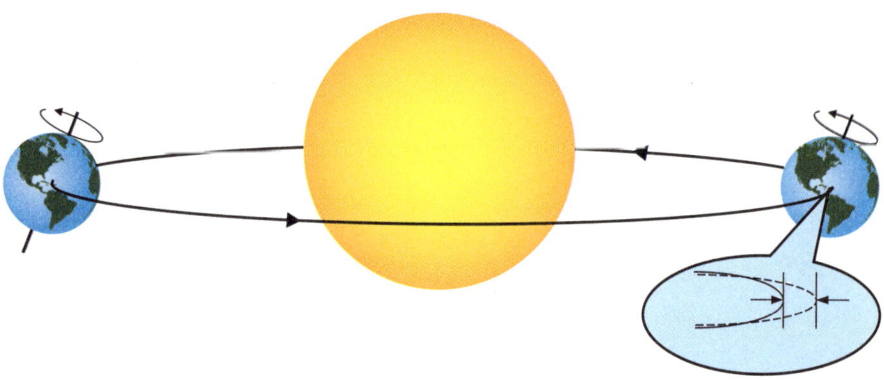

Figure D-1. Eccentricity of Earth's orbit around the sun.

The formula for eccentricity (e) is as follows:

$$e = \frac{\sqrt{a^2 - b^2}}{a}$$

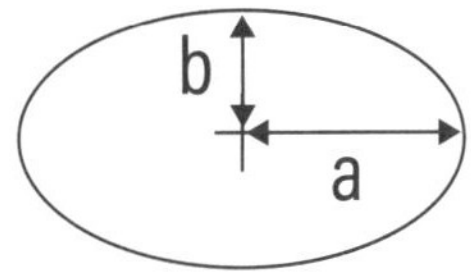

"a" and "b" are the semi-major and semi-minor ellipse axis distances, respectively. The diagram to the right of the equation, above, is an exaggerated illustration of an elliptical orbit. The current value of the Earth's eccentricity is quite small—meaning that the shape of the Earth's orbit is almost circular.[2]

The distance from the Earth to the sun is currently 93 million miles.[3] So, the nearest (dmin) and furthest distances (dmax) of the Earth from the sun, known as perihelion and aphelion, are:

$$d_{min} = a \times (1 - e) \cong [93{,}000{,}000 \times (1 - 0.0167)] \cong 91{,}500{,}000 \text{ miles;}$$

$$d_{max} = a \times (1 + e) \cong [93{,}000{,}000 \times (1 + 0.0167)] \cong 94{,}500{,}000 \text{ miles}$$

In summary, the Earth's distance from the sun varies between 91.5 and 94.5 million miles—a difference of three million miles—due to the elliptical nature of its orbit around the sun. Currently, the Earth is near the middle distance of 93 million miles from the sun. And the next time that the Earth will again be at this middle distance will be roughly 100,000 years from now.

appendix E

Contributors to the Greenhouse Effect

Recall in Chapter 9, I said that the primary atmospheric gases, and H_2O and CO_2, in particular, react quite differently to outgoing terrestrial radiation. Specifically, H_2O is 95% of the contribution to the greenhouse effect (GHE), CO_2 is approximately 3.6% of the GHE, and the other trace greenhouse gases (GHGs) comprise the remaining 1.4%.[1] These percentages are attributable to both the quantity and efficiency of gases participating in the greenhouse effect.

1. The **quantity, or volume fraction, of the various greenhouse gases** residing in the atmosphere are quite different. H_2O, as a gas, varies from nearly 0% in desert regions to 4% in the equatorial tropics. This is 0–40,000 ppm. On average, worldwide, H_2O vapor is very high at 3% or 30,000 ppm because of the amount of water on the surface of our planet. In contrast, the next highest volume fraction is CO_2, currently at approximately 410 ppm. So, there are about 73 times more water vapor molecules than CO_2 molecules in the air. Globally, water vapor in the atmosphere is relatively

flat at roughly 3%, while CO_2 is rising slowly and linearly at a rate of approximately 3 ppm per year.

2. The **efficiency of the various greenhouse gases** to absorb terrestrial radiation also favors water vapor to CO_2 due to the spectroscopic absorption properties of water vapor to CO_2.[2]

Figure E–1 illustrates the spectral emission band of terrestrial radiation, and the corresponding absorption bands of H_2O and CO_2.

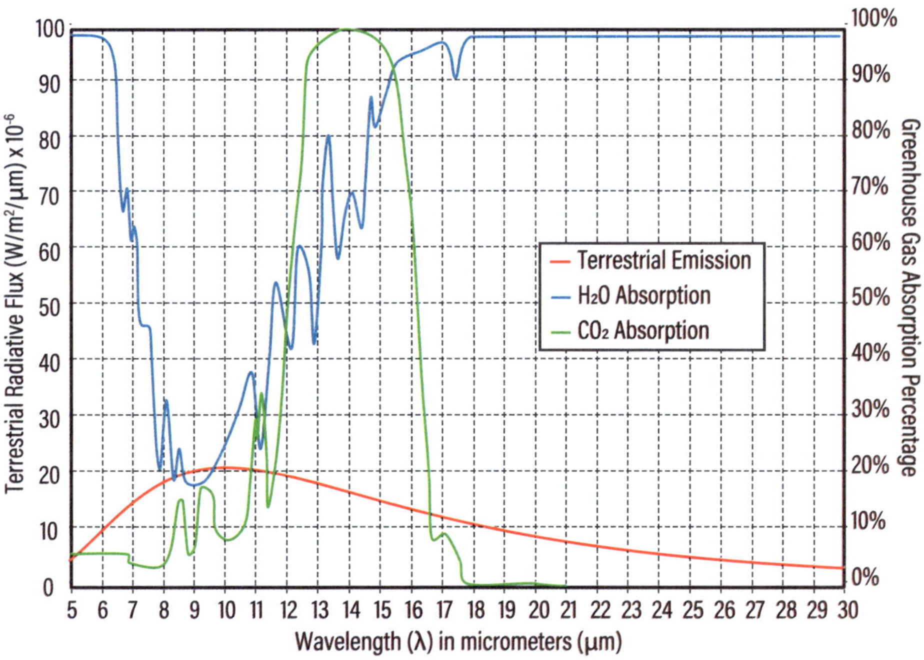

Figure E-1. Spectral emission of terrestrial radiation, and corresponding absorption of H_2O and CO_2.

Figure E–1 shows that absorption of terrestrial radiation is favorable to water vapor at all wavelengths except 12 to 15.5 μm, and then only slightly so.

The integrated product of emission, absorption, and volume fraction result in the greenhouse contribution factors of 95% for H_2O vapor, 3.6% for CO_2 gas, and 1.4% for the rest of the trace greenhouse gases combined.

Sources

Author's Note

[1] Middleton, David (2019). "Climate models have not improved in 50 years." *Watts Up With That.* December 4, 2019. Retrieved from: https://wattsupwiththat.com/2019/12/06/climate-models-have-not-improved-in–50-years/

Chapter 1. A Lifelong Interest

[1] Albert Einstein Quotes. Retrieved from: https://www.brainyquote.com/

[2] Crockford, Susan J. (2019). *The Polar Bear Catastrophe That Never Happened.* The Global Warming Policy Foundation.

[3] Christy, J. and Spencer, R. (2019). *Global Temperature Report: October 2019.* Earth System Science Center. University of Alabama—Huntsville. Retrieved from: https://www.nsstc.uah.edu/climate/2019/november2019/GTR_201911Nov_1.pdf

[4] Morice, C. P. et al (2012), *The HADCRUT4 Dataset,* Journal of Geophysical Research, 117.

Chapter 2. The Scorecard

[1] Baltimore Bulletin, *The Sporting Life.* 6 October 1894.

[2] Albert Einstein Quotes. Retrieved from: https://www.brainyquote.com/

[3] Christy, J. and Spencer, R. (2019). *Global Temperature Report: November 2019.* Earth System Science Center. University of Alabama—Huntsville. Retrieved from: https://www.nsstc.uah.edu/climate/2019/november2019/GTR_201911Nov_1.pdf/

Chapter 3. How Scientists Determine Truth

[1] Rowe, M., Facebook response to a viewer comment, regarding his narration of "How the Universe Works", Discovery Channel, Season 6, Episode 1, January 9, 2018.

[2] Smith, L., Chairman, Committee on Science, Space, and Technology. Full hearing on "Climate Science: Assumptions, Policy Implications, and the Scientific Method," U.S. House of Representatives, 29 March 2017.

[3] Summary of "Hypothesis vs. Theory" and "Theory vs. Law" criteria. *Diffen.com*. Diffen LLC, n.d. Retrieved from: https://www.diffen.com

[4] Russo, Remigio (1996). *Mathematical Problems in Elasticity, Vol 18.* World Scientific, p. 125.

[5] Israel, Hans; Ruckhaber, Erich; Weinmann, Rudolf, editors (1931). *Hundert Autoren gegen Einstein.* Leipzig: Voightländer.

[6] Hawking, Stephen (1998). *A Brief History of Time (10th Ed).* Bantam Books, p. 193.

[7] Merriam-Webster Learners Dictionary. An Encyclopedia Britannica Company. Retrieved from: http://www.learnersdictionary.com/definition/consensus

[8] Oreskes, N., *Beyond the Scientific Consensus on Climate Change.* Science, 306, p. 1686. June 2004.

[9] "Benny Peiser's Correspondence with Science." 7 May 2005. Retrieved from: *https://web.archive.org/web/20050507004218/http://www.staff.livjm.ac.uk/spsbpeis/Scienceletter.htm*

[10] Peiser, B., "The Dangers of Consensus Science." *National Post*. 17 May 2005.

[11] "Peiser's analysis of Oreske's abstracts." 4 July 2006. Retrieved from: *https://web.archive.org/web/20060704183216/http://www.staff.livjm.ac.uk/spsbpeis/Oreskes-abstracts.htm*

[12] Doran, P. and Zimmerman, M. "Examining the Scientific Consensus on Climate Change." *EOS, Transactions, American Geophysical Union*, Vol 90, Issue 3. P. 22–23. 20 January 2009.

[13] Anderegg, W. et al, "Expert Credibility In Climate Change." *Proceedings of the National Academy of Science.* 107(27). 12107–12109. 2010

[14] Cook, J., et al. (2013). "Quantifying the consensus on anthropogenic global warming in scientific literature." *Environmental Research Letters.* 8 024024 (7pp).

[15] Legates, D. R. (2013). "Climate consensus and 'misinformation': A rejoinder to agnotology, scientific consensus, and the teaching and learning of climate change." Science & Education 24:299–318.

[16] "97% study falsely classifies scientists papers' according to the scientists that published them." Popular Technology.net. May 21, 2013. Retrieved from: http://www.populartechnology.net/2013/05/97-study-falsely-classifies-scientists.html

[17] Stenhouse, N., et al (2014) "Meteorologists views about global warming: A survey of American Meteorological Society members." *Bulletin of the American Meteorological Society*. 95:1029–1040. Boston, MA.

[18] Maibach, E., Seitter, K., et al (2016). "A 2016 National Survey of American Meteorological Society Member Views on Climate Change: Initial Findings." *George Mason University, Center for Climate Change Communication*, Fairfax, VA.

[19] Hollande, M. F., *Paris Climate Conference—Opening speech by M. Francois Hollande*, Le Bourget, France, 30 November 2015.

[20] *Pope Francis, in an hour-long press conference on the plane that brought him back to Rome from a trip to Africa*. November 30, 2015.

[21] *M. Laurant Fabius, French Minister of Foreign Affairs and International Development, during a speech to the National Assembly*, Paris, France, 14 May 2014.

[22] *President Barack Obama climate change speech at Georgetown University*. Washington, DC, June 25, 2013.

[23] *Confirmation hearing for Dr. John Holdren, nominated to be Director of the Office of Science and Technology Policy in the Executive Office of the President*. U.S. Senate Commerce, Science, and Transportation Committee; February 12, 2009.

[24] Prince Charles of Wales speech. *Less Than 100 Months To Act*. Rio de Janeiro, Brazil, March 12, 2009.

[25] Reiss, R. *The Coming Storm*. Interview with James Hansen of NASA/GISS in 1988, predicting New York City's Westside Highway (which runs along the Hudson River) will be under water in 40 years (2028).

[26] John F. Kennedy. *Commencement Address At Yale University*. Jun 11, 1962

Chapter 4. The Big Picture—Looking Beyond What We Can See

[1] Zuber, M., "Science and Engineering Indicators 2018." *National Science Foundation, National Science Board*. Alexandria, VA, January 2018, Table 7-1, page 7-44.

[2] NASA Goddard Space Flight Center. "Earth's Atmospheric Layers." Greenbelt, Maryland, May 2015, Retrieved from: https://www.nasa.gov/image-feature/goddard/earths-atmospheric-layers/

[3] NASA Glenn Research Center. "Earth's Atmospheric Layers." Cleveland, Ohio, May 2015, Retrieved from: https://www.grc.nasa.gov/WWW/K–12/airplane/atmosmet.html/

[4] NASA, NOAA, and the USAF (1976). *U.S. Standard Atmosphere, 1976.* Pg. 23. Washington, DC. Retrieved from: https://ntrs.nasa.gov/archive/nasa/casi.ntrs.nasa.gov/19770009539.pdf

Chapter 5. The Really Big Picture—In Space

[1] NASA Jet Propulsion Laboratory. "Our Solar System." Pasadena, California, January 2018. Retrieved from: https://solarsystem.nasa.gov/solar-system/our-solar-system/in-depth/

[2] NASA Goddard Space Flight Center (2018). *Sun Fact Sheet, Sun/Earth Comparison.* Retrieved from: https://nssdc.gsfc.nasa.gov/planetary/factsheet/sunfact.html/

[3] NASA Goddard Space Flight Center (2018). *Earth Fact Sheet.* Retrieved from https://nssdc.gsfc.nasa.gov/planetary/factsheet/earthfact.html/

[4] C. Q. Choi (2018). "Planet Earth: About Its Orbit, Atmosphere & Size." Science & Astronomy, Space.com. October 10, 2018. Retrieved from: https://www.space.com/54-earth-history-composition-and-atmosphere.html

[5] "Resolution B3 on recommended nominal conversion constants for selected solar and planetary properties." International Astronomical Union. 2015. Retrieved from: www.iau.org/static/resolutions/IAU2015_English.pdf

[6] Lindsey, Rebecca (2009). "Climate and Earth's energy budget (p11)." *NASA Goddard Space Flight Center—Earth Observatory.* Greenbelt, MD. Jan 14, 2009. Retrieved from: https://earthobservatory.nasa.gov/features/EnergyBalance

[7] "Sunspot data from the World Data Center—Sunspot Index and Long Term Solar Observations (WDC-SILSO)." *Royal Observatory of Belgium.*

[8] G. Kopp, et al, "The impact of the revised sunspot record on solar irradiance reconstructions." *Solar Physics.* 2016.

[9] G. Kopp, "Earth's incoming energy; the total solar irradiance," in Reference Module in Earth Systems and Environmental Sciences, Elsevier, 2016.

[10] *Fundamentals of Physics.* David Halliday and Robert Resnick. John Wiley and Sons, Inc., 1974.

[11] *The Cambridge Planetary Handbook.* Michael Bakich, Cambridge University Press, 2000. Pg. 49.

[12] NASA Goddard Space Flight Center, Earth Observatory (2000). Obliquity. Retrieved from: https://earthobservatory.nasa.gov/features/Milankovitch/milankovitch_2.php/

[13] NASA Goddard Space Flight Center (2017). *Earth Fact Sheet, Orbital parameters.* Retrieved from https://nssdc.gsfc.nasa.gov/planetary/factsheet/earthfact.html/

[14] R. Weast and M. Astle, Editors. "Physical data (eccentricity) for the planets, their satellites, and some asteroids." *Handbook of Chemistry and Physics, 62nd Edition.* CRC Press, Boca Raton, Florida. 1981–1982. Pg. F–141.

[15] A. Berger and M. F. Loutre. "Insolation values for the climate of the last 10 million years." Quaternary Science Reviews, 10: 297–317. 1991.

[16] J.L. Simon et al (1994). "Numerical expressions for precession formulae and mean elements for the moon and planets." *Astronomy and Astrophysics.* 282 (2).

[17] Windows to the Universe team (2008). *Milankovitch Cycles.* Retrieved from https://www.windows2universe.org/?page=/earth/climate/cli_sun.html

[18] S. Khan (2011). *Milankovitch cycles precession and obliquity—Cosmology & Astronomy.* Khan Academy. Mountain View, California. Retrieved from: https://youtu.be/ZD8THEz18gc.

[19] C. R. Scotese (2015). *Phanerozoic Temperature Curve.* PALEOMAP Project. Evanston, IL Retrieved from: https://www.academia.edu/12114306/Phanerozoic_Global_Temperature_Curve

Chapter 6. The Really Big Picture—In Time

[1] NASA Goddard Institute for Space Studies (2018). *Worldwide Temperature Measuring Stations by Latitude and Longitude.* New York City, NY. Retrieved from: https://data.giss.nasa.gov/gistemp/stdata/

[2] R. Spencer. (2016). Measuring Global Temperatures: Satellites or Thermometers. Retrieved from: http://www.cfact.org/2016/01/26/measuring-global-temperatures-satellites-or-thermometers/

[3] NOAA Satellite Information System (2018). *NOAA's Geostationary and Polar-Orbiting Weather Satellites.* Retrieved from: https://noaasis.noaa.gov/NOAASIS/ml/genlsatl.html/

[4] W. Ferguson. (2013). *Ice Core Data Help Solve a Global Warming Mystery.* Scientific American. Retrieved from: https://www.scientificamerican.com/article/ice-core-data-help-solve/

[5] C. Scotese. (2002). *Temperature oscillations in Geological Eras.* Retrieved from: https://www.geocraft.com/WVFossils/CO2_Temp_02.html

[6] C. R. Scotese (2015). *Phanerozoic Temperature Curve.* PALEOMAP Project. Evanston, IL Retrieved from: https://www.academia.edu/12114306/Phanerozoic_Global_Temperature_Curve

[7] Christy, J. and Spencer, R. (2019). *Global Temperature Report: November 2019.* Earth System Science Center. University of Alabama—Huntsville. Retrieved from: https://www.nsstc.uah.edu/climate/2019/november2019/GTR_201911Nov_1.pdf

[8] C. Morice (2019). HadCRUT4 Near Surface Global Temperature Dataset. Met Office, Hadley Centre. Exeter, UK. Retrieved from: https://www.metoffice.gov.uk/hadobs/hadcrut4/data/current/download.html

[9] Jouzel, J., et al (2007). "EPICA Dome C Ice Core 800 KYr Deuterium Data and Temperature Estimates." *NOAA/NCDC Paleoclimatology Program.* Boulder CO. Retrieved from: ftp://ftp.ncdc.noaa.gov/pub/data/paleo/icecore/antarctica/epica_domec/edc3deuttemp2007.txt

[10] A. Gasparrini, et al. "Mortality risk attributable to high and low ambient temperature: a multi-country observational study." *The Lancet.* Volume 386, pp 369–375. London, UK. July 25, 2015. Retrieved from: https://www.thelancet.com/pdfs/journals/lancet/PIIS0140-6736(14)62114-0.pdf

[11] C. R. Scotese (2015). *Phanerozoic Temperature Curve.* PALEOMAP Project. Evanston, IL Retrieved from: https://www.academia.edu/12114306/Phanerozoic_Global_Temperature_Curve

[12] Ruddiman, W. F. (2001). *Earth's climate—past and future.* W. H. Freeman & Sons. New York, NY.

[13] NOAA National Centers for Environmental Information, World Data Center for Paleoclimatology (2019). *Paleoclimatology Data.* Boulder, CO.

Chapter 7. The Air We Breathe

[1] Murphy, B. (2019). "Photograph of Rural Northern Utah."

[2] NASA, Goddard Space Flight Center (2017). *Earth Fact Sheet, Terrestrial Atmosphere.* Retrieved from: https://nssdc.gsfc.nasa.gov/planetary/factsheet/earthfact.html/

[3] Utah Department of Environmental Quality (2019). *Air Quality Trend Charts.* Salt Lake City, UT. Retrieved from: https://air.utah.gov/trendcharts.php?id=slc

[4] U.S. EPA (2019). Air quality trends for ozone. Retrieved from: https://www.airnow.gov/index.cfm?action=pubs.aqiguideozone

5 M. McElroy (2002). *The Atmospheric Environment.* Princeton University Press.

6 U.S. EPA (2019). Outdoor Air Quality Data, Air Data—Concentration Plots. Retrieved from: https://www.epa.gov/outdoor-air-quality-data/air-data-concentration-plot.

7 World Meteorological Organization (2019). Surface Synoptic Observations, Division of Air Quality. Geneva, Switzerland. Retrieved from: https://public.wmo.int/en

Chapter 8. The "Bad" Guy and "Good" Guy Cases: CO_2 and H_2O

1 Patrick Moore (2015). "Should we celebrate carbon dioxide?" Presentation at: *The 2015 Global Warming Policy Foundation (GWPF) annual meeting.* London, UK. October 14, 2015. Retrieved from: https://youtu.be/57W3ZhOAkAE

2 Dana Royer (2006). "CO_2-forced climate thresholds during the Phanerozoic." Geochimica et Cosmochimica Acta. 70 (2006), pp 5665–5675. Retrieved from: www.sciencedirect.com

3 Dieter Lüthi, et al (2008). " High-resolution carbon dioxide concentration record 650,000—800,000 years before present." *Nature.* Volume 453, pp. 379–382. May 15, 2008. Retrieved from: EPICA Dome C Ice Core 800KYr Carbon Dioxide Data, NOAA NCEI Paleoclimatology Program, Boulder, CO.

4 Lindsey, Rebecca (2018). "Climate change: Atmospheric carbon dioxide." *NOAA Climate.gov News & Features, Understanding Climate, National Centers for Environmental Information (NCEI).* Boulder, CO. August 1, 2018. Retrieved from: https://www.climate.gov/news-features/understanding-climate/climate-change-atmospheric-carbon-dioxide

5 Humlum, Ole (2013). "The phase relation between atmospheric carbon dioxide and global temperature." *Global and Planetary Change. Elsevier, Science Direct.* Volume 100, January 2013. Pp 51–69. Amsterdam, The Netherlands. Retrieved from: https://www.sciencedirect.com/science/article/pii/S0921818112001658.

6 MacRae, Allan (2019). "CO_2, global warming, climate and energy." *Watts Up With That.* June 15, 2019. Retrieved from: https://wattsupwiththat.com/2019/06/15/co2-global-warming-climate-and-energy–2/

7 Jeremy Beaman (2018). "Alabama's state climatologist John Christy debunks this month's climate hysterics." *Climate Change Dispatch.* August 16, 2018. Retrieved from: https://climatechangedispatch.com/?s=christy

[8] Astronaut Walt Cunningham (2017). "CO_2 levels in spacecraft." *Twelfth International Conference on Climate Change.* Sponsored by The Heartland Institute, Chicago, IL. March 23, 2017. Retrieved from: https://youtu.be/reVmhKjH_M8

[9] "Distribution of Earth's Water." *U.S. Geological Survey.* December 2, 2016. Retrieved from: https://water.usgs.gov/edu/earthwherewater.html

[10] Petsch, S. (2014). "The global oxygen cycle." *Treatise on geochemistry (2nd Ed).* 10. Pp 437–474.

[11] Walker, J.C. (1980). *The oxygen cycle in the natural environment and the biogeochemical cycles.* Springer-Verlag, Berlin, FRG.

[12] Preston Cloud and Aharon Gibor (1970). "The oxygen cycle." *Scientific American.* 223:3, pp 110–123. September 1970.

[13] S. B. Idso and B. A. Kimball (1994). "Effects atmospheric CO_2 enrichment on biomass accumulation and distribution in Eldarica pine trees." *Journal of Experimental Botany* **45**: 1669–1672. Retrieved from: https://plantsneedco2.org/default.aspx?menuitemid=327

[14] Howard Perlman (2016). "How much water is there on, in, and above the Earth." *U.S. Geological Survey.* Reston, VA. December 2, 2016. Retrieved from: https://water.usgs.gov/edu/earthhowmuch.html and https://water.usgs.gov/edu/earthwherewater.html

[15] Lindsey, Rebecca (2009). "Climate and Earth's energy budget." NASA Goddard Space Flight Center—Earth Observatory. Greenbelt, MD. January 14, 2009. Retrieved from: https://earthobservatory.nasa.gov/features/EnergyBalance

Chapter 9. Energy Balance And The Greenhouse Effect

[1] Lindsey, Rebecca (2009). "Climate and Earth's energy budget (p13)." *NASA Goddard Space Flight Center—Earth Observatory.* Greenbelt, MD. Jan 14, 2009. Retrieved from: https://earthobservatory.nasa.gov/features/EnergyBalance

[2] Lindsey, Rebecca (2009). "Climate and Earth's energy budget (p7)." *NASA Goddard Space Flight Center—Earth Observatory.* Greenbelt, MD. Jan 14, 2009. Retrieved from: https://earthobservatory.nasa.gov/features/EnergyBalance

[3] Lindsey, Rebecca (2009). "Climate and Earth's energy budget (p2)." *NASA Goddard Space Flight Center—Earth Observatory.* Greenbelt, MD. Jan 14, 2009. Retrieved from: https://earthobservatory.nasa.gov/features/EnergyBalance

[4] "Global warming: A closer look at the numbers." Retrieved from: https://www.geocraft.com/WVFossils/greenhouse_data.html

[5] Michaels, Patrick (1998). "Global deception: The exaggeration of the global warming threat." *Center for the Study of American Business, Washington University* **168,** pg. 8. St. Louis, MO. June 1998. Retrieved from: https://www.geocraft.com/WVFossils/Reference_Docs/PMichaels_Jun98.pdf

[6] Stephens, Graeme, et al (2012). "An update on Earth's energy balance in light of the latest global observations." *Nature Geoscience* **5**, 691–696. September 23, 2012. Retrieved from: https://www.nature.com/articles/ngeo1580

[7] Cortez, Alexandria Ocasio (D-NY). "The world's gonna end in 12 years if we don't address climate change." Conversation with Ta-Nehisi Coates at the fourth annual MLK NOW event in New York City. Sarah Sloat reporting for: Inverse.com. January 22, 2019. Retrieved from: https://www.inverse.com/article/52659-alexandria-ocasio-cortez-climate-change

[8] Nye, Bill (2019). "AOC gets it . . . We can address climate change together and get to work. Twitter, #SXSW@AOC. March 9, 2009.

[9] Hawking, Stephen (2018). "Stephen Hawking was terrified of climate change." Report by Ari Phillips of *Earther*. March 4, 2018. Retrieved from: https://earther.gizmodo.com/stephen-hawking-was-terrified-of-climate-change-1823761657

[10] Hansen, James (1988). "The greenhouse effect: Impacts on current global temperature and regional heat waves." Statement by Hansen to: *The United States Senate Committee On Energy and Natural Resources*. Washington, DC. January 23, 1988. Retrieved from: https://climatechange.procon.org/sourcefiles/1988_Hansen_Senate_Testimony.pdf

[11] Scotese, Christopher (2015). *Phanerozoic Temperature Curve*. PALEOMAP Project. Evanston, IL Retrieved from https://www.academia.edu/12114306/Phanerozoic_Global_Temperature_Curve

[12] Jouzel, J., et al (2007). "EPICA Dome C Ice Core 800 KYr Deuterium Data and Temperature Estimates." *NOAA/NCDC Paleoclimatology Program*. Boulder CO. Retrieved from: ftp://ftp.ncdc.noaa.gov/pub/data/paleo/icecore/antarctica/epica_domec/edc3deuttemp2007.txt

[13] Berner, Robert & Kothavala, Zavareth (2001). "GEOCARB III: A revised model of atmospheric CO_2 over Phanerozoic time." American Journal of Science 301, 182–204. February 1, 2001. Retrieved from: https://www.geocraft.com/WVFossils/Reference_Docs/Geocarb_III-Berner.pdf

[14] Giaever, Ivar (2015). "Global warming revisited—Where the temperature is measured." Lecture by Nobel Laureate Ivar Giaever presented at: *The 65th Lindau Nobel Laureate Meeting*. Lindau, Bavaria, Germany. July 1,2015. Original NASA GISS GISSTEMP graph as captured from the video of Giaever's lecture at: https://youtu.be/TCy_UOjEir0 (5:10–6:40 minute segment)

[15] Obama, Barack (2015). "No challenge presents a greater threat to future generations than climate change." *Sixth presidential state of the union address.* Washington, DC. January 20, 2015. Retrieved from: https://www.azquotes.com/author/11023-Barack_Obama/tag/climate-change

[16] Giaever, Ivar (2015). "Global warming revisited—Response to President Obama warning about climate change." Lecture by Nobel Laureate Ivar Giaever presented at: *The 65th Lindau Nobel Laureate Meeting.* Lindau, Bavaria, Germany. July 1,2015. Original NASA GISS GISSTEMP graph as captured from the video of Giaever's lecture at: https://youtu.be/TCy_UOjEir0 (25:40–26:40 minute segment)

Chapter 10. Tornadoes

[1] Kathryn Prociv (2013). "From domestic to international: Tornadoes around the world." *Tornado Climatology.* July 25, 2013. Retrieved from: http://www.ustornadoes.com/2013/07/25/from-domestic-to-international-tornadoes-around-the-world/

[2] NOAA National Weather Service, Storm Prediction Center (2019). *U.S. Tornadoes (1950–2018).* Norman, OK. Retrieved from: https://www.spc.noaa.gov/wcm/data/1950–2017_all_tornadoes.csv

[3] NOAA National Weather Service, Storm Prediction Center (2019). *The EnhancedFujita Scale (EF Scale).* Norman, OK. Retrieved from: https://www.spc.noaa.gov/efscale/

[4] NOAA National Weather Service, Storm Prediction Center (2019). *U.S. Tornadoes (1950–2018) Sorted By EF Scale.* Norman, OK. Retrieved from: https://www.spc.noaa.gov/wcm/data/1950–2017_all_tornadoes.csv

Chapter 11. Tropical Hurricanes, Typhoons, and Cyclones

[1] Gentry, R. C., (1970). "Project Stormfury Hurricane Debbie modification experiments, August 1969." *Science* **168**, 473–475. Retrieved from: https://www.researchgate.net/publication/249614939_Project_STORMFURY_A_Scientific_Chronicle_1962–1983

[2] Underwood, L.J., CDR, USN (1969). "Flight pattern—eyewall experiment Flight 'F' (Navy WC-121N) Altitude 1,000 ft inflow monitor." *Tab D to Appendix I to Annex E of Project Stormfury Operation Plan No. 1–69.* Retrieved from: https://apps.dtic.mil/dtic/tr/fulltext/u2/722670.pdf

3 Willoughby, H.e., et al (1985). "Project Stormfury: A scientific chronicle 1962–1983." *Bulletin of the American Meteorological Society* **66**(5), 505–514. Retrieved from: https://www.researchgate.net/publication/249614939_Project_STORMFURY_A_Scientific_Chronicle_1962-1983

4 Knapp, K. R., et al (2010). "The International Best Track Archive for Climate Stewardship (IBTrACS)." *Bulletin of the American Meteorological Society*, **91**, 363–376. Retrieved from: http://tropical.atmos.colostate.edu/Realtime/index.php?arch&loc=northatlantic

5 NOAA National Hurricane Center (2012). *The Saffir-Simpson Hurricane Wind Scale*. Miami, Florida. Retrieved from: https://www.nhc.noaa.gov/pdf/sshws.pdf

Chapter 12. More Extreme Weather-Related Events

1 Fefferman, C. (2000). "Existence and smoothness of the Navier-Stokes Equation(s)." *Clay Mathematics Institute*. Cambridge, Massachusetts. Retrieved from: http://www.claymath.org/sites/default/files/navierstokes.pdf

2 Andela, N., et al (2017). "A human-driven decline in global burned area." Science, 356:6345, pp 1356–1362. June 30, 2017. Retrieved from: http://science.sciencemag.org/content/356/6345/1356

3 "Total wildland fires and acres (1926–2018)." *National interagency fire center*. Boise, ID. Retrieved from: https://www.nifc.gov/fireInfo/fireInfo_stats_totalFires.html

4 Mass, Cliff (2018). "Was global warming a significant factor in California's Camp Fire? The answer is clearly no." *Cliff Mass weather and climate blog*. November 20, 2018. Retrieved from: https://cliffmass.blogspot.com/2018/11/was-global-warming-significant-factor.html

5 Hughes, Mimi, et al (2009). "Anthropogenic reduction of Santa Ana winds." California climate change center. Sacramento, CA. Retrieved from: https://pdfs.semanticscholar.org/e334/ca53cb872ca92b242fb9e69223b42e611bb5.pdf

6 Christy, J. and Spencer, R. (2019). *Global Temperature Report: November 2019*. Earth System Science Center. University of Alabama—Huntsville. Retrieved from: https://www.nsstc.uah.edu/climate/2019/november2019/GTR_201911Nov_1.pdf

7 "Total wildland fires and acres (1926–2018)." *National interagency fire center*. Boise, ID. Retrieved from: https://www.nifc.gov/fireInfo/fireInfo_stats_totalFires.html

8 "Lightning and human-caused fires." *National interagency fire center*. Boise, ID. Retrieved from: https://www.nifc.gov/fireInfo/fireInfo_stats_lightng-human.html

[9] Kummer, Larry (2018). "Wildfires and climate change: fake news in action." Fabius Maximus website. November 21, 2018. Retrieved from: https://fabiusmaximus.com/2018/11/21/wildfires-the-latest-extreme-climate-change/

[10] U.S. Environmental Protection Agency (2016). "Climate change indicators: High and low temperatures." Retrieved from: https://www.epa.gov/climate-indicators/climate-change-indicators-high-and-low-temperatures#6

[11] NOAA National Centers for Environmental Information (2019). "Historical Palmer drought indices." Boulder, CO. Retrieved from: https://www.ncdc.noaa.gov/temp-and-precip/drought/historical-palmers/

[12] Gasparrini, A., et al (2015). "Mortality risk attributable to high and low ambient temperature: a multi-country observational study." *The Lancet.* Volume 386, pp 369–375. London, UK. July 25, 2015. Retrieved from: https://www.thelancet.com/pdfs/journals/lancet/PIIS0140–6736(14)62114–0.pdf

[13] NOAA Center for Operational Oceanographic Products and Services—CO-OPS (2019). "U.S. linear relative sea level trends and 95% confidence limits in mm/yr and ft/century." Silver Spring, MD. Retrieved from: https://tidesandcurrents.noaa.gov/sltrends/mslUSTrendsTable.html

[14] Permanent Service for Mean Sea Level (2019). "Global Tide Gauge Data." *UK National Oceanography Centre.* Liverpool, UK. Retrieved from: https://www.psmsl.org/data/

[15] Noerdlinger, P., and Brower, K. (2007). "The melting of floating ice raises the ocean level." *Geophysical Journal International.* 170:1, 145–150. July 1, 2007. Retrieved from: https://academic.oup.com/gji/article/170/1/145/2019346

[16] Fleming, Kevin et al (1998). "Refining the eustatic sea-level curve since the last glacial maximum using far and intermediate field sites." Earth and planetary science letters **163** (1–4): pp 327–342. Retrieved from: https://commons.wikimedia.org/wiki/File:Post-Glacial_Sea_Level.png

[17] Murphy, Zoeann and Mooney, Chris (2019). "Gone in a generation: Across America, climate change is already disrupting lives." *The Washington Post.* January 29, 2019. Retrieved from: https://www.washingtonpost.com/graphics/2019/national/gone-in-a-generation/flooding-climate-change.html?utm_term=.bf0c2d1a80b3

[18] "Floods: Recurrence intervals and 100-year floods (USGS)." *U.S. Geological Survey.* Retrieved from: https://water.usgs.gov/edu/100yearflood.html

[19] Klotzbach, P., and Bell, M. (2018). "Qualitative discussion of Atlantic Basin Seasonal Hurricane Activity for 2019." Department of Atmospheric Science, Colorado State University, Fort Collins, CO. December 13, 2018. Retrieved from: https://tropical.colostate.edu/media/sites/111/2018/12/2018–12.pdf

[20] Reed, Andra, et al (2015). "Increased threat of tropical cyclones and coastal flooding to New York City during the anthropogenic era." *Proceedings of the National Academy of Sciences.* **112** (41), pp 12,610–12,615. October 13, 2015. Retrieved from: https://www.pnas.org/content/112/41/12610

[21] Chu, Jennifer (2012). "With climate change, today's '100-year floods may happen every three to 20 years: research." Massachusetts Institute of Technology. Retrieved from: https://phys.org/news/2012-02-climate-today-year-years.html

[22] Grossman, Lisa (2015). "Floods that hit New York City every 500 years now hit every 24." New Scientist—Earth. September 28, 2015. Retrieved from: https://www.newscientist.com/article/dn28254-floods-that-hit-new-york-city-every-500-years-now-hit-every-24/

[23] Goldenberg, Suzanne (2013). "Climate study predicts a watery future for New York, Boston, and Miami." The Guardian. July 29, 2013. Retrieved from: https://www.theguardian.com/environment/2013/jul/29/climate-new-york-boston-miami-sea-level

Chapter 13. Summary

[1] Goreham, Steve. (2017). "Climate science and myths of renewable energy." Presentation to the *Friends of Science.* Ottawa, Canada. July 1, 2017. Retrieved from: https://www.youtube.com/watch?v=mtHreJbr2WM (at the 4–5 minute mark of this video presentation)

[2] NOAA National Centers for Environmental Information (2019). "U.S. Historical Climatology Network (USHCN), land-based station datasets." Ashville, NC. Retrieved from: https://www.ncdc.noaa.gov/cdo-web/search

[3] C. Morice (2019). HadCRUT4 Near Surface Global Temperature Dataset. Met Office, Hadley Centre. Exeter, UK. Retrieved from: https://www.metoffice.gov.uk/hadobs/hadcrut4/data/current/download.html

Appendix A: Apple-Earth Comparison

[1] NASA Goddard Space Flight Center (2018). *Earth Fact Sheet.* Retrieved from https://nssdc.gsfc.nasa.gov/planetary/factsheet/earthfact.html/

[2] Von Kármán, Theodore (1967). *The Wind and Beyond*, p. 343.

Appendix B: Sun Power

[1] International Astronomical Union. "Resolution B3 on recommended nominal conversion constants for selected solar and planetary properties." 2015. Retrieved from: https://www.iau.org/static/resolutions/IAU2015_English.pdf

[2] NASA Goddard Space Flight Center (2018). *Sun Fact Sheet.* Retrieved from https://nssdc.gsfc.nasa.gov/planetary/factsheet/sunfact.html/

[3] NASA Goddard Space Flight Center (2018). *Earth Fact Sheet.* Retrieved from https://nssdc.gsfc.nasa.gov/planetary/factsheet/earthfact.html/

[4] Lindsey, Rebecca (2009). "Climate and Earth's energy budget (p11)." *NASA Goddard Space Flight Center—Earth Observatory.* Greenbelt, MD. Jan 14, 2009. Retrieved from: https://earthobservatory.nasa.gov/features/EnergyBalance

Appendix C: Gravity

[1] *Fundamentals of Physics.* David Halliday and Robert Resnick. John Wiley and Sons, Inc., 1974.

Appendix D: Earth's Orbital Eccentricity

[1] R. Weast and M. Astle, Editors. "Physical data for the planets, their satellites, and some asteroids." *Handbook of Chemistry and Physics, 62nd Edition.* CRC Press, Boca Raton, Florida. 1981–1982. Pg. F–141.

[2] A. Berger and M. F. Loutre. "Insolation values for the climate of the last 10 million years." Quaternary Science Reviews, 10: 297–317. 1991.

[3] NASA Goddard Space Flight Center (2018). *Sun Fact Sheet, Sun/Earth Comparison.*

Appendix E: Contributors to the Greenhouse Effect

[1] Michaels, P. (1998). "Global Deception: The Exaggeration of the Global Warming Threat." *Center for the Study of American Business,* Policy Study Number 146. Washington University, St. Louis, Missouri. June 1998. Pg. 8.

[2] Rothman L. S., et al (2004). "The HITRAN molecular spectroscopic database." Journal of Quantitative Spectroscopy & Radiative Transfer 96: 139–204.

About the Author

Bill Pekny is a scientist, engineer, and program manager. His fifty-year career has covered twenty-two years as an active duty, regular Army officer; five years of government civil service; nineteen years in the aerospace industry; and four years independently researching and writing this, his first book. He has experienced the life cycle of several laser, nuclear, and meteorological projects for the U.S. military and aerospace industry.

Chronological highlights of his military and aerospace industry careers include:

- Research meteorologist assignments with the U.S. Navy Weather Research Facility and Weather Reconnaissance Squadron Four—the Navy's renowned "Hurricane Hunters."
- Research meteorologist assignments developing aerosol diffusion models and radioactive fallout prediction tools for the U.S. Army.
- Chemical officer with the U.S. Army's 25th Light Infantry Division.
- Nuclear design physicist at the Lawrence Livermore National Laboratory.
- Visiting Scholar at Stanford University's Ginzton Laboratory of Applied Physics.
- Laser physicist and avionics project manager for the U.S. Army's Aviation Program Executive Office.
- Laser physicist and deputy program manager for the development, testing, production, and deployment of infrared lasers for

military applications. Bill and his team received two U.S. patents for their laser design efforts.

- Program manager for the deactivation of the "Peacekeeper" Intercontinental Ballistic Missile.
- Systems engineer for "Minuteman III" missile sustainability and refurbishment.

The thread running through much of Bill's work was the knowledge of what is going on in the air, as applied to military and aerospace applications.

Bill holds a Bachelor of Science degree in physics from DePaul University, as well as a Master of Science Degree in physics from the Georgia Institute of Technology (Georgia Tech). He also completed a combined total of three additional years of graduate coursework in physical meteorology and numerical analysis at the Florida State University and the University of Utah.

Bill is an accomplished technical writer and speaker. And now, he also is a published author who hopes that everyone will gain and enjoy a greater understanding of our incredible Earth-atmosphere system and its miraculous ability to nourish all life on this planet.

Bill resides in rural northern Utah, USA. He is the husband of forty-nine years to eternal partner Nancy, father of three superb sons, two delightful daughters-in-law, and three glorious grandchildren.

Acknowledgments

I wish to thank the many hands that have contributed immensely to the preparation of this book.

My heartfelt thanks first go to my immediate family, especially my wife and "ghost" contributor, Nancy, sons Brian, Matt, and Dan, and their families, as well as Nancy's sister Shawna and her husband Roy Knorr. They have provided key inputs and encouragement throughout the research, writing, and review of this book. You make everything I do achievable.

Thanks to my good parents Hank and Martha, members of The Greatest Generation, who made sure I had a good education and a love for learning. Thanks also to my brother, Bob, who excelled academically, and motivated me to do the same.

A huge thanks goes out to my editor, Susan Suffes. I deeply appreciate this New York City lady's expert guidance and patience as we completed and polished the book. She provided inestimable value throughout the process.

Barry Kerrigan, Del Lemond, and the staff at Desktop Miracles, Inc., have been incredibly helpful to me, a first-time author, in the understanding, development, and execution of the publishing process. They have been patient, constructive, kind, and meticulous throughout the effort.

Many thanks go to Dr. Steven Saterlie, PhD in physics and long time friend and associate in the aerospace industry. We bounced many thoughts and ideas off each other over the years. I would never have completed this book without that interaction, or your timely, thoughtful, and detailed scientific review of the entire book.

Dr. Jerry Simons, doctorate in meteorology and industrial electronics, was there for me every step of the way as an experienced scientific advisor, a sounding board for ideas, and strategist for the marketing of the book. Thank you, Jerry, for your much needed encouragement, technical feedback, and business acumen.

Looking back at my own education and subsequent workforce experience, these are a several prominent educators and associates who made science fun for me. Let's address them separately.

First are the patient, talented, and compassionate educators that molded me into the scientist that I am today:

L.C. Haley, M. R. Barry, M. R. Kelleher, and J. C. Crane—high school mathematics and physics teachers at Brother Rice High School in Oak Lawn, Illinois.

Drs. Tony Behof, Mary Boas, and Louis Aquila—undergraduate level physics and mathematics professors at DePaul University, Chicago, Illinois.

Dr. Ward Seguin—graduate level marine meteorology research advisor at Florida State University, Tallahassee, Florida.

Drs. Mark Christensen, Charles Braden, Donald O'Shea, and Earl McDaniel—graduate level professors of mathematics and physics at the Georgia Institute of Technology, Atlanta, Georgia.

Drs. W. R. Gearhart and S. K. Kao—graduate level professors of mathematics and meteorology at the University of Utah, Salt Lake City, Utah.

Drs. Bob Byer, Tony Siegman, and Marty Fejer—postgraduate level professors and mentors at Stanford University's Ginzton Laboratory of Applied Physics, Stanford, California.

Many coworkers and mentors helped me steadily mature as a scientist, which ultimately enabled me to write this book.

Drs. Mike Anastasio, Dan Huval, Harry Vantine, Cory Call, Jay Chase, Mike Nelson, Peter Newcomb, Jim Routh, Gordon Repp, Bob Poor, and Richard Couch—physics colleagues at Lawrence Livermore National Laboratory, Livermore, California.

Art Zabinski, Ron Langietti, Paul Lang, and Drs. Mark Christensen, Jerry Griffith, Hao Tran, and Jim Messina—physics mentors and colleagues in the aerospace industry.

Dr. Matt Mikulich—Friend, colleague, and oil industry geophysicist.

Jerry Nickerson and Drs. R. Cecil Gentry and Robert H. Simpson—hurricane meteorology mentors during Project Stormfury. Dr. Simpson co-developed the renowned Saffir-Simpson Hurricane Wind Scale.

And last, but certainly not least, I also am grateful to the wide variety of people that reviewed the book for its readability. These are friends, family, and associates that range from students and teachers to an assortment of white and blue-collar professionals:

Ella Simons
Kathy Krebs
Roger & Julie Pyle
Colleen and Gary Lemon
Dr. Richard & Mary Moutvic
Dr. Richard Curtis
Dr. Kaye Nichols
Dr. Dean Hughes
Dr. Chris Wasden
Allyson Ollivier
Nyal, Bette, & Glen McMullin
Brent & LeAnn Hathaway
Gordon & Jean Jolley
Judge Stuart Waldrip
Dr. Matthew, Merry, Kyle, & Kira Evitts

Your thoughts, inspiration, encouragement, and constructive criticism have made this book a reality. I thank you, one and all, for your generous contributions.

Index

Figures are indicated by page numbers followed by *f*.

C

D

E

F

G

H

I

J

K

L

M

N

O

P

R

S

T

U

V

W

Z